MICROWAVE STUDIES OF EXCITON CONDENSATION IN GERMANIUM

ISSLEDOVANIE KONDENSATSII EKSITONOV V GERMANII METODAMI SVCH

ИССЛЕДОВАНИЕ КОНДЕНСАЦИИ ЭКСИТОНОВ В ГЕРМАНИИ МЕТОДАМИ СВЧ

The Lebedev Physics Institute Series

Editors: Academicians D. V. Skobel'tsyn and N. G. Basov

P. N. Lebedev Physics Institute, Academy of Sciences of the USSR

Recent Volumes in this Series

Volume 57	Theory of Interaction of Elementary Particles at High Energies
Volume 58	Investigations in Nonlinear Optics and Hyperacoustics
Volume 59	Luminescence and Nonlinear Optics
Volume 60	Spectroscopy of Laser Crystals with Ionic Structure
Volume 61	Theory of Plasmas
Volume 62	Methods in Stellar Atmosphere and Interplanetary Plasma Research
Volume 63	Nuclear Reactions and Interaction of Neutrons and Matter
Volume 64	Primary Cosmic Radiation
Volume 65	Stellarators
Volume 66	Theory of Collective Particle Acceleration and Relativistic Electron Beam Emission
Volume 67	Physical Investigations in Strong Magnetic Fields
Volume 68	Radiative Recombination in Semiconducting Crystals
Volume 69	Nuclear Reactions and Charged-Particle Accelerators
Volume 70	Group-Theoretical Methods in Physics
Volume 71	Photonuclear and Photomesic Processes
Volume 72	Physical Acoustics and Optics: Molecular Scattering of Light; Propagation of Hypersound; Metal Optics
Volume 73	Microwave—Plasma Interactions
Volume 74	Neutral Current Sheets in Plasmas
Volume 75	Optical Properties of Semiconductors
Volume 76	Lasers and Their Applications
Volume 77	Radio, Submillimeter, and X-Ray Telescopes
Volume 78	Research in Molecular Laser Plasmas
Volume 79	Luminescence Centers in Crystals
Volume 80	Synchrotron Radiation
Volume 81	Pulse Gas-Discharge Atomic and Molecular Lasers
Volume 82	Electronic Characteristics and Electron—Phonon Interaction in Superconducting Metals and Alloys
Volume 83	Theoretical Problems in the Spectroscopy and Gas Dynamics of Lasers
Volume 84	Temporal Characteristics of Laser Pulses and Interaction of Laser Radiation with Matter
Volume 85	High-Power Lasers and Laser Plasmas
Volume 86	Superconductivity
Volume 87	Coherent Cooperative Phenomena
Volume 88	Cosmic Rays in the Stratosphere and in Near Space
Volume 89	Electrical and Optical Properties of III—IV Semiconductors
Volume 90	The Kinetics of Simple Models in the Theory of Oscillations
Volume 91	Lasers and Their Applications in Physical Research
Volume 93	Techniques and Methods of Radio-Astronomic Reception
Volume 94	Pulsed Neutron Research
Volume 96	Problems in the General Theory of Relativity and Theory of Group Representations
Volume 97	Excitons and Domain Luminescence of Semiconductors
Volume 100	Microwave Studies of Exciton Condensation in Germanium

In preparation:
Volume 99	Stimulated Raman Scattering

Proceedings (Trudy) of the P. N. Lebedev Physics Institute

Volume 100

Microwave Studies of Exciton Condensation in Germanium

Edited by
N. G. Basov

P. N. Lebedev Physics Institute
Academy of Sciences of the USSR
Moscow, USSR

Translated from Russian by
David L. Burdick

CONSULTANTS BUREAU
NEW YORK AND LONDON

Library of Congress Cataloging in Publication Data

Main entry under title:

Microwave studies of exciton condensation in germanium.

(Proceedings (Trudy) of the P. N. Lebedev Physics Institute; v. 100)
Translation of Issledovanie kondensatsii éksitonov v germanii metodami svch.
Includes indexes.
1. Exciton theory. 2. Germanium — Spectra. 3. Microwave spectroscopy. I. Basov, Nikolaĭ
Gennadievich, 1922-   II. Series: Akademiia nauk SSSR. Fizicheskiĭ institut. Proceedings;
v. 100.
QC1.A4114 vol. 100 [QC176.8.E9] [530.4'1]   530'.08s   78-10160
ISBN 978-1-4615-8569-5   ISBN 978-1-4615-8567-1   (eBook)
DOI 10.1007/978-1-4615-8567-1

The original Russian text was published by Nauka Press in Moscow in 1977 for the Academy
of Sciences of the USSR as Volume 100 of the Proceedings of the P. N. Lebedev Physics
Institute. This translation is published under an agreement with the Copyright Agency of
the USSR (VAAP).

Proceedings (Trudy) of the P. N. Lebedev Physics Institute

Volume 100

Microwave Studies of Exciton Condensation in Germanium

Edited by
N. G. Basov

P. N. Lebedev Physics Institute
Academy of Sciences of the USSR
Moscow, USSR

Translated from Russian by
David L. Burdick

CONSULTANTS BUREAU
NEW YORK AND LONDON

Library of Congress Cataloging in Publication Data

Main entry under title:

Microwave studies of exciton condensation in germanium.

(Proceedings (Trudy) of the P. N. Lebedev Physics Institute; v. 100)
Translation of Issledovanie kondensatsii éksitonov v germanii metodami svch.
Includes indexes.
1. Exciton theory. 2. Germanium – Spectra. 3. Microwave spectroscopy. I. Basov, Nikolaĭ
Gennadievich, 1922-   II. Series: Akademiia nauk SSSR. Fizicheskiĭ institut. Proceedings;
v. 100.
QC1.A4114 vol. 100 [QC176.8.E9] [530.4′1]   530′.08s   78-10160
ISBN 978-1-4615-8569-5   ISBN 978-1-4615-8567-1   (eBook)
DOI 10.1007/978-1-4615-8567-1

The original Russian text was published by Nauka Press in Moscow in 1977 for the Academy of Sciences of the USSR as Volume 100 of the Proceedings of the P. N. Lebedev Physics Institute. This translation is published under an agreement with the Copyright Agency of the USSR (VAAP).

PREFACE

This volume examines a timely problem in solid state physics, the condensation of excitons in semiconductors. The papers in this volume describe a new approach to the study of exciton condensation based on microwave-frequency conductivity and the breakdown of an exciton gas in a microwave field, an effect discovered by the authors. These methods have been used in extensive studies of germanium and have produced important new data about the condensation process. In addition to the experimental data, a theory is presented for the microwave breakdown of excitons in the presence of electron—hole droplets. Also included are three articles devoted to the experimental aspects of low-temperature semiconductor research. This volume will interest scientists and engineers involved with research in solid state physics.

CONTENTS

Introduction   1
     A. A. Manenkov

Microwave Studies of Exciton Condensation in Ge   3
     G. N. Mikhailova

Microwave Breakdown of Excitons and the Kinetics of Free Carriers and
    Excitons in Germanium in the Presence of Electron-Hole Drops 37
     A. A. Manenkov, V. A. Milyaev, G. N. Mikhailova,
       V. A. Sanina, and A. S. Seferov

Microwave Breakdown Studies of Exciton Condensation in Ge and Luminescence
    during One- and Two-Photon Excitation of Carriers 45
     G. V. Zubov, A. A. Manenkov, V. A. Milyaev, G. N. Mikhailova,
       T. M. Murina, A. M. Prokhorov, and A. S. Seferov

Microwave Absorption by Nonequilibrium Current Carriers in Germanium.
    A Method for Determining Carrier Concentration 51
     B. V. Zubov, A. A. Manenkov, V. A. Milyaev, G. N. Mikhailova,
       T. M. Murina, V. A. Sanina, and A. S. Seferov

Microwave Methods for Studying Exciton Condensation in Semiconductors 59
     A. A. Manenkov

A Helium-3 Refrigerator for Studying Excitons in Ge at Temperatures
    below 1°K 67
     S. I. Valyanskii, V. A. Milyaev, G. N. Mikhailova,
       and A. B. Fradkov

Effects of Laser Heating on Germanium Samples at Liquid Helium Temperature 77
     A. A. Manenkov, G. N. Mikhailova, A. S. Seferov
       and V. D. Chernetskii

Cyclotron Resonance and Radiative Recombination in Pure Ge with Laser Excitation . . 87
     V. P. Aksenov, N. B. Volkov, B. G. Zhurkin,
       and I. G. Maksimchuk

INTRODUCTION

A.A. Manenkov

Recently a new phenomenon in semiconductor physics has been the object of much interest. It is the condensation of excitons into electron-hole droplets, a process which occurs when there is a high density of nonequilibrium carriers generated by some means (such as optical excitation). This phenomenon was predicted by L. V. Keldysh [1] from an analysis of experimental data on photoconductivity in Ge. The prediction was derived from theoretical considerations developed in analogy with the behavior of alkali metals. The prediction was soon verified experimentally by Pokrovskii and Svistunova [2], who observed a new line in the recombination radiation spectrum of Ge in addition to the usual exciton line. The new line is due to recombination of carriers inside the droplet. This work was followed by intensive investigations of the condensation phenomenon using a variety of methods, among which optical studies must be mentioned (luminescence, light scattering, absorption of infrared radiation, and so on). There have also been studies of some microwave properties (cyclotron resonance and Alfvén resonance) and acoustic properties (ultrasonic absorption) in Ge in the presence of condensed excitons. These studies produced interesting data concerning the condensation process and provided many characteristics of the condensate itself (density and size of the electron-hole droplet, phase diagram parameters, and so on). A discussion of these results is found in [3].

Recently a series of papers by Jeffries and co-workers has excited great interest. They report observations of large (0.3 mm diameter) droplets in germanium formed when the sample is in nonuniform compression. The existence of such drops was demonstrated by photographs obtained by converting the infrared radiation from the drop to visible light. The existence of large drops has also been confirmed in infrared absorption and scattering experiments [4] and microwave conductivity measurements [5].

This volume is a compilation of work performed at the Oscillations Laboratory of the P. N. Lebedev Physics Institute, Academy of Sciences of the USSR, from 1971 to 1975. The thrust of these studies is the use of microwave methods to investigate exciton condensation in Ge. More specifically, we have studied the microwave conductivity of Ge crystals excited by pulsed optical radiation and the microwave breakdown of excitons in the presence of electron-hole droplets. The breakdown effect was first reported by Manenkov et al. [6] and is a very informative approach to many of the droplet parameters and the properties and characteristics of the entire three-component system in the presence of condensation (free carriers, free excitons, and electron-hole droplets). The microwave method* provides a rich storehouse of qualitative and quantitative data about exciton condensation and enables one to determine many

---

* A discussion of this method and references to the original articles are found in the first four articles of this volume.

parameters which are found to be in good agreement with data obtained by other methods and with the theoretical model. We should also mention that, because of its simplicity as an experimental approach and with respect to theoretical interpretation, these microwave methods result in reliable information which significantly extends and improves the results obtained by other techniques.

The first four articles of this volume deal with microwave breakdown studies of excitons in germanium. The first three describe in detail the experimental studies of exciton condensation under various conditions, present a theory for the microwave breakdown of excitons in the presence of droplets, and interpret the observations and some of the quantitative data derived from those interpretations concerning the combined three-component system of carriers in germanium. The fourth article presents additional analysis of the microwave data using the aforementioned theoretical model which describes the kinetics of a two-phase system (exciton gas and droplet) in both weak and strong microwave electric fields. It also presents a table of all the fundamental parameters of the system of carriers, excitons, and droplets obtained by comparing the theory with experiment.

These four articles are followed by two papers devoted to the experimental methods for studying semiconductors at low temperatures with strong optical excitation (a description of a He^3 refrigerator for reaching temperatures below 1°K and a study of the effects of sample heating by the laser radiation). There is also an article describing the microwave measurement of the density of nonequilibrium carriers in semiconductors. This paper should be of great interest to all those dealing with the physics of nonequilibrium processes in semiconductors.

LITERATURE CITED

1. L. V. Keldysh, Proceedings of the Ninth International Conference on Semiconductor Physics [in Russian], Nauka, Moscow (1969), p. 1384.
2. Ya. E. Pokrovskii and K. I. Svistunova, Pis'ma Zh. Éksp. Teor. Fiz., 9:435 (1969).
3. Excitons in Semiconductors [in Russian], Nauka, Moscow (1971); Ja. Pokrovsky, Phys. Status Solidi (A), 11:385 (1972); C. D. Jeffries, Science, 189:955 (1975).
4. Ya. E. Pokrovskii and K. I. Svistunova, Pis'ma Zh. Éksp. Teor. Fiz., 23:110 (1976).
5. A. A. Manenkov, V. A. Milyaev, G. N. Mikhailova, and A. S. Seferov, Pis'ma Zh. Éksp. Teor. Fiz., 24:141 (1976).
6. A. A. Manenkov, V. A. Milyaev, G. N. Mikhailova, and S. P. Stolin, Pis'ma Zh. Éksp. Teor. Fiz. 16:454 (1972).

MICROWAVE STUDIES OF EXCITON CONDENSATION IN Ge*

G. N. Mikhailova

The breakdown of excitons in Ge by a 10-GHz microwave field is studied in detail at helium temperatures in both the continuous and pulsed modes. The dependence of the threshold breakdown power on the duration and time delay of the microwave pulse relative to the laser pulse is investigated together with its dependence on the level of optical excitation at 1.3°K. We present a theory for the effect based on the collision ionization of excitons in equilibrium with electron-hole droplets. The breakdown effect is also discussed from the biexciton point of view. The experimental data are in good agreement with the droplet model. The radius, density, and lifetime of electron-hole drops are obtained.

INTRODUCTION

The number of studies of excitons in semiconductors, especially those concerned with collective motions, has greatly increased in recent years. To a large extent such work has been stimulated by the availability of lasers which make it rather easy to produce large concentrations of nonequilibrium carriers in samples.

Excitons in semiconductors are characterized by large effective radii (many times greater than the lattice constants) and small binding energies as compared with atomic energies. As a result interactions between excitons become important even at low concentrations, and collective effects result from such interactions.

Questions concerning the collective properties of excitons first arose in theoretical studies of the possibility of Bose condensation in the exciton system, and in the related phenomena of superfluidity and superconductivity [1-3]. It was demonstrated that Bose condensation was possible when the predominant interaction between excitons was repulsive, and that excitonic molecules could be formed if the predominant interaction were attractive [4, 5].

Using theoretical models developed in analogy with alkali metals, Keldysh [6-8] proposed yet another possibility, that of exciton condensation into a liquid phase (metallic drop). Such a change in an exciton system should have all the characteristics of a Type 1 phase transition, and the resulting dense electron-hole phase must be similar to a liquid metal.

Experimentally one can study the exciton system over a wide range of concentrations ranging from the gaseous state to that of a degenerate electron-hole plasma. The most inter-

---
* Based on a dissertation submitted for the degree of Candidate of Physical and Mathematical Science and defended in 1974 at the P. N. Lebedev Physics Institute.

esting region is that of intermediate concentrations where the condition $(n a_0^3)^{1/3} \sim 1$ is satisfied (n is the concentration and a_0 the effective exciton radius). Although one can compute exactly the energy spectrum of the equilibrium system at both large ($n a_0^3 \gg 1$) and small ($n a_0^3 \ll 1$) exciton concentrations, the problem is much more difficult at intermediate densities because there is no small parameter in terms of which expansions can be carried out.

The majority of the experimental work on condensed excitons has employed germanium because Ge is the material best suited to verification of theoretical condensation models. Germanium has been studied extensively, is available in very pure form, and has a band structure in which indirect excitons have a rather long lifetime $\sim 10^{-5}$ sec at low temperatures.

Currently there are two schools of thought on the behavior of exciton systems in Ge as the density increases. At the time this paper was begun the question concerning the nature of exciton complexes in Ge was still very much undecided. Some authors have maintained [9-11] that as the density increases, the excitons bind into molecules (biexcitons) much like gaseous hydrogen, and with a further in increase density and in drop temperature the biexciton gas begins to liquefy. At the same time others [12-17] assumed that the exciton gas undergoes a direct transition from the gas to the liquid metal state or electron-hole droplet, minimizing the biexciton stage. They suggest that electron-hole droplets are formed even at very low densities of the exciton gas when the temperature is appropriate.

Various experimental approaches have been used to study these systems; recombination radiation is most commonly used, but constant-current photoconductivity, absorption and emission in the far IR, and scattering of IR radiation have also been employed. In addition, the effects of external factors such as magnetic and electric fields, deformation, and others, have been used to study exciton complexes.

The condensed state of excitons can be studied in detail using microwave methods such as microwave conductivity and cyclotron and size resonance. Microwave probes have a number of advantages. First, microwave conductivity is a highly sensitive direct method for observing carriers over a wide range of densities, including very low density. Microwave experiments avoid undesirable surface and contact effects, including carrier injection into the sample. Finally, by using microwave cavities one can study the effects of either electric or magnetic fields, separately, on excitons and exciton formation processes.

We have performed microwave conductivity studies on pure Ge excited by a laser. During the course of our studies we made the first observations of excitons in Ge using continuous microwave fields [18]. To obtain a better understanding of the nature of the condensed state we studied the effect of a pulsed microwave field on excitons [19, 20]. The dependence of the breakdown threshold on the duration and time delay between the microwave pulse and laser pulse and on the laser intensity enabled us to analyze in detail the exciton breakdown process and to suggest a theory for the effect* based on collision ionization of excitons in equilibrium with electron-hole droplets. Thus, the breakup of excitons by pulsed microwave fields is a new and effective method for studying the collective properties of excitons.

1. COLLECTIVE PROPERTIES OF EXCITONS IN SEMICONDUCTORS

1.1. Theoretical Work. Excitonic Molecules

(Biexcitons) and Electron-Hole Droplets

An exciton is formed because of the Coulomb interaction between an electron and a hole. In semiconductors having large dielectric constants and low effective masses for both electrons

---

* The theory of microwave breakdown of excitons in the presence of electron-hole droplets was developed by L. V. Keldysh.

and holes, excitons have radii which are more than an order of magnitude greater than the lattice constant. This means that excitons can be viewed as hydrogen-like impurities in the semiconductor, and the effective mass approximation (Wannier model) can be used.

The binding energy $\mathcal{E}_0$ of an exciton and its effective radius a_0 can be calculated using the Bohr model for atomic hydrogen if one recognize that the Coulomb interaction between the electron and hole with effective masses m_e and m_h is reduced because of polarization in the crystal [8]:

$$\mathcal{E}_0 = \frac{1}{2}\frac{e^4 m}{\varepsilon_0^2 \hbar^2}, \qquad a_0 = \frac{\varepsilon_0 \hbar^2}{m e^2},$$

where $m = m_e m_h / (m_e + m_h)$ is the reduced mass of the electron and hole, and ε_0 is the dielectric constant. The binding energy and exciton radius in germanium are ($\varepsilon_0 = 16$; $m_e = 0.082 m_0$, $m_h = 0.34 m_0$) $\mathcal{E}_0 \simeq 3.5$ meV and $a_0 \simeq 120$ Å according to these equations.

Excitons are either direct or indirect, depending on the band structure of the semiconductor. For direct excitons the energy minimum is found at $\mathbf{k} = 0$. An indirect exciton results whenever either the maximum of the valence band or the minimum of the conduction band is found where $\mathbf{k} \neq 0$.

The greatest difference between direct and indirect excitons is in their respective lifetimes relative to radiative recombination. Since an indirect exciton can radiatively recombine only with the emission or absorption of a phonon, the rate of radiative recombination for indirect excitons is two to three orders of magnitude slower than that of direct excitons. At low temperatures ($kT \ll \mathcal{E}_0$) the lifetime of a direct exciton is $\sim 10^{-7}$-10^{-9} sec while that of an indirect exciton is $\sim 10^{-4}$-10^{-6} sec.

At low temperatures ($kT \ll \mathcal{E}_0$) and concentrations such that $n a_0^3 \ll 1$ the interactions between excitons and with the lattice can be treated as weak. Thus, we can treat a low density system of excitons as an ideal Boltzmann gas of quasiatoms.

As the density n increases, the distance between excitons becomes comparable to their Bohr radius, and the kinetic energy of an electron or hole (the Fermi energy) $\hbar^2 n^{2/3}/m = 2\mathcal{E}_0 \times (n a_0^3)^{2/3}$ becomes comparable to their average potential energy for interactions with other particles $e^2 n^{1/3}/\varepsilon_0$. In the high-density limit ($n a_0^3 \gg 1$) the kinetic energy is much larger than the interaction energy, and excitons can no longer be formed. Calculations show that at such densities the Coulomb potential is screened at distances large compared to the exciton Bohr radius. At this point the electrons and holes act like a degenerate Fermi gas [21].

We shall be interested in the intermediate density region where the parameter $(n a_0^3)^{1/3}$ is of the order of unity. We will examine some of the collective effects in a system of excitons, effects due to interactions between excitons.

If we view an exciton as a "single-electron atom" then, as a first approximation, we can treat the behavior of excitons as their density increases in analogy with hydrogen atoms or alkali metal atoms. As the density of a gas of any matter increases, the attraction between the atoms or molecules causes condensation. However, we know that there are differences between the condensation of hydrogen atoms and alkali metal atoms. Hydrogen condenses into a molecular liquid which is a dielectric, while alkali metals condense into atomic liquids which are metallic. The great difference in the behavior of these two substances is explained by noting that the binding energy of a hydrogen molecule is much larger than that of a molecule of an alkali metal.

One would expect that a system of excitons would condense in a somewhat similar manner as the density increases. There are thus two possibilities. In the first, the excitons first form molecules (biexcitons) and then condense into a dense liquid as the concentration increases still further. The second possibility is that the excitons condense into a metallic phase and essentially minimize the biexciton phase. The biexciton-molecule route was suggested by Lampert [4] and Moskalenko [5] in 1958. Based on positronium molecule calculations [22] Lampert estimated that the dissociation energy for the biexciton should be about $\mathscr{E}_m \sim 0.1\mathscr{E}_0$, where $\mathscr{E}_0$ is the exciton binding energy.

An exact calculation of the biexciton binding energy poses some difficulty. One cannot use the adiabatic approximation (useful for hydrogen molecules) because the biexciton consists of four light particles. A variational method can be used to solve the problem. Since in many semiconductors the effective masses of electrons and holes are not equal, one must calculate the biexciton binding energy as a function of the ratio $\sigma = m_e/m_l$. Sharma [23] was the first to attempt to solve this problem. He calculated $\mathscr{E}_m$ as a function of σ between the limits $\sigma \approx 0$ (the hydrogen molecule) and $\sigma = 1$ (the positron molecule). His results indicate that the biexciton is unstable when $0.2 < \sigma < 0.4$. Subsequently, however, Akimoto and Hanamura [24] discovered an error in Sharma's work. They then computed the binding energy for $0 < \sigma < 1$ for electrons and holes with isotropic masses. They found that the biexciton binding energy decreases monotonically with σ and where $\sigma = 1$, $\mathscr{E}_m = 0.018 \mathscr{E}_P$ ($\mathscr{E}_P$ is the positron Rydberg).

Brinkman et al. [25] improved the calculation of $\mathscr{E}_m$. First they considered the isotropic-masses case and found that in the limit $(\sigma = 1)$ $\mathscr{E}_m = 0.029 \mathscr{E}_P$. Then they attempted to take into account the anisotropy of the valence band and calculated the biexciton binding energy for CuCl, CuBr, Cu_2O, and other semiconductors. For Ge their calculations with anisotropic electron masses ($m_{e\parallel}/m_{e\perp} = 0.052$) showed that in Ge the biexciton binding energy is $\mathscr{E}_m \sim 0.1$ meV. As in the positronium molecule the absence of heavy particles (such as the protons in hydrogen molecules) in the biexciton causes large-amplitude null vibrations (of the order of the exciton radius) which tend to loosen the molecular binding [8].

Because the binding energy of the biexciton is so small it is natural to expect that the behavior of an exciton system will be more like that of the alkali metal gas than of a hydrogen gas. This analogy has been used [6-8] to construct a qualitative model of the transition of an exciton gas into a condensed phase and to predict the fundamental properties of the condensate and the features of its formation. According to the predictions of this model, as the exciton density increases at temperatures below the critical temperature, one would expect the exciton gas to go into a liquid phase and to form electron-hole droplets. Below the critical temperature the carrier concentration in the droplet will be some orders of magnitude greater than the density of the exciton gas in equilibrium with the drops. In analogy with the alkali metals it was shown in [8] that the critical temperature for such a transition T_c will be determined by the condition $kT_c \sim 0.1\mathscr{E}_0$. For Ge $\mathscr{E}_0 \simeq 4$ meV and $T_c \simeq 5°K$; therefore the critical concentration n_c at which the evaporation temperature of the drop reaches its maximum value of T_c can be found from the condition $n_c a_0 \sim 10^{-1}-10^{-2}$ ($n_c \sim 10^{15}-10^{16}$ cm^{-3}). The particle density n_0 in the drop was estimated from the condition $n_0 a_0 \sim 1$, which gives $n_0 \simeq 10^{17}$ cm^{-3} for Ge.

The key to the question as to the form in which excitons exist at high densities is found in the binding energy of the particles in the drop. In order for electron-hole drops to evolve from an exciton gas it is necessary that the energy E_c of a pair of particles in the condensate be lower than the energy of an exciton. If this condition is not fulfilled there will be no condensation; instead, at high densities ($na_0^3 \sim 1$) a Mott transition [26] from the dielectric gas to an electron-hole plasma will take place.

Let us consider the particle energy in a drop in more detail. Pokrovsky et al. [27, 28] have used a simple model to show that in the condensed phase the plasma energy is the sum of the energies of the electron and hole gases. The total internal energy of the drop can therefore be written as

$$E_C = \mathcal{E}_F^e + \mathcal{E}_F^h + \mathcal{E}_{ex}^{ee} + \mathcal{E}_{ex}^{hh} + \mathcal{E}_{corr}^{ee} + \mathcal{E}_{corr}^{hh} + \mathcal{E}_{corr}^{eh},$$

where $(\mathcal{E}_F^e + \mathcal{E}_F^h)$ is the sum of the Fermi energies of the electrons and holes, and

$$(\mathcal{E}_{ex}^{ee} + \mathcal{E}_{ex}^{hh}) \quad \text{and} \quad (\mathcal{E}_{corr}^{ee} + \mathcal{E}_{corr}^{hh} + \mathcal{E}_{corr}^{eh})$$

are the sums of the exchange and correlation energies for both electrons and holes. The first terms in this expression (the kinetic and exchange energies) can be calculated exactly from the germanium band structure.

It is very difficult to determine $\mathcal{E}_{corr}$ because it can be computed for just the two limiting cases $na_0^3 \ll 1$ and $na_0^3 \gg 1$. The Wigner expression has been used in an interpolation approach [27, 28] to calculate $\mathcal{E}_{corr}^{ee} + \mathcal{E}_{corr}^{hh}$ in the case where $na_0^3 \sim 1$; they assumed that $\mathcal{E}_{corr}^{eh}$ was not smaller than either $\mathcal{E}_{corr}^{hh}$ or $\mathcal{E}_{corr}^{ee}$. The equilibrium density n_0 of carriers in the drop can be determined from the minimum-energy condition $\partial E_C / \partial n |_{n=n_0} = 0$. By using this model for Ge it was found that when the charge carrier concentration in the electron-hole drop was $2 \cdot 10^{17}$ cm^{-3} the energy per pair of particles in the drop was 2 meV lower than in the exciton.

More exact calculations of the ground state of an electron-hole plasma have been performed by Brinkman et al. [29] using a modified random phase approximation (RPA). They demonstrated that an electron-hole liquid is unstable in an ideal isotropic band structure of a semiconductor. The following assumptions were made for Ge: the conduction band was approximated by four spherical subbands, the valence band by two, and no distinction was made between light and heavy holes. They then found that in Ge the metallic phase is more energetically favorable than the exciton gas, and there is an energy minimum E_C when $n_0 = 1.8 \cdot 10^{17}$ cm^{-3}, which is 2 meV lower than $\mathcal{E}_0$. Thus, it is possible to form a condensate in Ge because there are four equivalent energy minima, but condensation is not possible in crystals with a single minimum.

Finally, Combescot and Nozieres [30] calculated the ground state of the electron-hole plasma by including the entire Ge band structure except the warping of the valence band. They found that in this electron-hole gas there is a minimum energy E_C which is 2.5 meV below the exciton ground level $\mathcal{E}_0 = -3.6$ meV corresponding to a carrier concentration of $\sim 2 \cdot 10^{17}$ cm^{-3}. They pointed out the excellent agreement of their calculated values for n_0 and E_C with experimental values and estimated their results to be accurate to within 10%.

As a consequence, theory predicts the following pattern for the transition from the exciton gas to the condensed state in Ge at low temperatures. When the density is low ($\sim 10^{10}$-10^{11} cm^{-3}) most of the carriers are in excitons and only a small fraction are in biexcitons, if they are formed at all. As the crystal passes through the critical density there appear regions with increased concentrations of electrons and holes. These are the electron-hole drops whose dimensions increase with the carrier density (as do their numbers) even though the exciton density is unchanged. The carrier density in the drops is constant and independent of temperature. The transition from excitons to an electron-hole liquid is similar to a Type I phase transition. The absence of heavy particles in the condensed phase makes it impossible for them to exist in a crystalline state even at absolute zero temperature because the amplitude of their zero oscillations is comparable to the interparticle separation. In this respect the exciton system is similar to liquid helium which also does not solidify at ordinary pressures because of its zero-point oscillations.

1.2. Review of Experimental Work *

Studies of exciton condensation in semiconductors, particularly Ge and Si, are very extensive, and a great deal of progress has been achieved in understanding this effect. The funda-

---

* This review covers the work published up to the middle of 1975.

mental experimental methods for studying collective effects in exciton systems are luminescence [9, 10, 12, 17, 31, 32], constant-current photoconductivity [33-35], light scattering [36-
39], cyclotron resonance [15, 40-43], infrared absorption and emission [14, 44], and current
fluctuations in p—n junctions [45-47]. A number of papers are devoted to the effects of various external factors on exciton formation. These include uniform [13] and nonuniform [48, 49]
pressure, magnetic fields [50-53], and a constant electric field [54]. In our microwave studies
[18-20] of the conductivity of Ge we observed the breakup of excitons due to the microwave
field. This is very informative because it allows one to determine the fundamental characteristics of the exciton—electron-hole drop system.

The report by Markiewicz et al. [55] of the observation of large (up to 1 mm diameter)
drops in Ge under pressure has excited much interest. These drops were observed by luminescence and dimensional resonance. A photograph has been published [56] of a large drop. The
picture was obtained through conversion of infrared radiation into light.

In the present review we shall analyze in detail those experimental works which, in our
opinion, address the most basic questions concerning exciton condensation and which determine the most important parameters of an electron-hole drop. We shall only mention other
work.

Haynes [57] was the first to report observations of exciton complexes (1966). The recombination radiation spectrum of pure Si at T < 10°K showed a new line displaced to the long
wavelength side of the free exciton line at E = 15 meV. In view of the fact that the intensity
of this line increased as the square of the exciting radiation intensity, Haynes assigned the
line to emission from biexcitons. He proposed a recombination mechanism in which one of the
excitons forming the molecule recombines, giving off part of the energy of the other exciton
at the same time. The second exciton then decays into a free electron and hole. However,
subsequent experiments have shown that the dependence on the excitation intensity is not always
quadratic, and can vary over wide limits [58, 59]; thus it is not associated with the biexciton.

Asnin and Rogachev [33] observed a sharp increase in the photoconductivity of Ge at an
exciton density of $n = 10^{16}$ cm^{-3}. Although they erroneously interpreted this effect as a transition from a metal to a dielectric (a Mott transition) their work stimulated further theoretical
and experimental studies of exciton condensation.

The appearance of a new line in the recombination radiation spectrum of pure Ge was
first observed in the experiments of Pokrovskii and Svistunova [12, 28]. The line was about
5 meV below the line associated with the annihilation of a free exciton accompanied by the
emission of an LA phonon (714 meV). The authors estimated that the LA(709) line arose when
the concentration of electron-hole pairs was about 10^{14} cm^{-3} with T ≤ 2.8°K. The intensity of
the new line increased by about a factor of 100 over a very narrow temperature range, and the
range was very dependent on the excitation level. It was demonstrated that the quantum yield
of the LA(709) line was $\eta = 0.8-1$ and the de-excitation time was $\tau = 20$ μsec. A model was
suggested [28] for the condensed phase based on electron-hole drops which qualitatively explained the position and line width and the dependence of the line intensity on the excitation
level. Based on this model the equilibrium particle density in the condensed phase was estimated to be $n_0 = 2 \cdot 10^{17}$ cm^{-3} [60]. Later work has produced better results concerning the
shape and location of the LA(709) line and dependence on the excitation level. Resonance absorption and emission from Ge with a maximum at a photon energy of 8.7 meV have been reported [14, 44]. This effect has been interpreted as a plasma resonance in the metallic drop.
The important result derived from this work is the parameters characterizing the electron-
hole drop: the drop size is $R \simeq 10-20$ μm, the carrier plasma frequency is $\omega_{pl} \simeq 2 \cdot 10^{13}$ sec^{-1},
and the effective carrier collision frequency in the drop is $\nu \simeq 7 \cdot 10^{12}$ sec^{-1}.

Keldysh [8] has examined certain effects due to the high mobility of electron-hole drops. In particular, he has studied their motion in nonuniform fields. The high mobility arises because of the Fermi degeneracy, for the phonon scattering of electrons and holes comprising the drop is small compared with scattering of free carriers at the same temperature.

Experimental proof of the high mobility in the drops has been obtained in studies of radiative recombination in Ge under uniaxial compression [48, 49]. Alekseev et al. [48] directly observed the motion of both excitons and drops in the temperature range 4.2-1.8°K when Ge is deformed nonuniformly. It was found that the drops have a much higher mobility than the free excitons. A 4-mm displacement of drops was observed in that experiment.

Interesting results were obtained by studying the recombination radiation from Ge in ultrahigh magnetic fields [50-53]. Intensity oscillations in the LA(709) line have been reported [50]. These results were interpreted as oscillations in the electron Fermi energy due to Landau quantization of the degenerate plasma in the drop. By using the period in terms of the reciprocal of the field $\Delta(1/H)$ they found $\mathscr{E}_F^e$ for a sample oriented $\langle 100 \rangle$ together with the equilibrium carrier concentration $n_0 = p_0 = 2 \cdot 10^{17}$ cm^{-3}. They also have observed [51] the LA(709) line split into two lines in an external field of 40-70 kG (for the case $H \| \langle 100 \rangle$). These effects are easily understood in terms of a degenerate drop, and cannot be interpreted using the biexciton model.

Oscillations in the far infrared absorption spectrum of Ge have been observed [52]. The results were interpreted just as in [50] by using the electron-hole drop model.

Recently Betzler et al. [53] published some work on magnetooscillations in luminescence from electron-hole drops in Ge. They studied the kinetics and line shapes of the LA(709) and LA(729) lines in a 32-kG field at T = 1.5°K. Oscillations were observed in the luminescence intensity, lifetime, and line width of the drop emissions. The results were interpreted using the Keldysh–Silin theory [61] which predicts oscillations of the free energy of a drop in a magnetic field. This causes oscillations in the particle density in the electron-hole drop in an external magnetic field, leading to the observed effects. In addition, they determined the quantum yield of radiation by the drop (25%) and demonstrated that the main radiationless process in a drop is Auger recombination [53]. The probabilities for radiative recombination and Auger recombination were found to be B = $3 \cdot 10^{-14}$ cm^3/sec and C = $4 \cdot 10^{17}$ cm^6/sec respectively.

Elastic scattering of light proves to be an important new approach to the study of collective effects in a system of excitons. Since light scattering by biexcitons is negligible, scattering experiments were able to resolve the questions surrounding which model to use. Moreover these scattering experiments provided much valuable information concerning the properties of drops. Pokrovskii and Svistunova [36] were the first to observe light scattered by drops. They estimated the drop radii to be 3-8 μm at 2°K, and found the carrier concentrations in the drops to be $n_0 = 2 \cdot 10^{17}$ cm^{-3}.

Subsequently, many others [37, 38, 62] conducted detailed studies of electron-hole drops using scattered light. It was shown that as the temperature changed from 1.9 to 4.2°K the drop radius increased from 3 to 12 μm, and the densities varied between 10^8 and $3 \cdot 10^3$ cm^{-3} depending on the temperature and method and intensity of the excitation. It was observed that the drop size increased and the concentration decreased as the temperature increased. New data on the condensation kinetics have been obtained. It was shown that when T $\leq$ 3°K the formation of incipient liquid phases is related to fluctuations in exciton density, and that at higher temperatures the concentration of drops is determined by the number of condensation centers. The surface tension has been found to be $1.6 \cdot 10^{-4}$ erg/cm^2 for drops in Ge [62].

There are a number of studies of the kinetics for recombination radiation [10, 31, 63, 64]. Zubov et al. [10, 63] first performed two-photon bulk excitation of carriers in Ge by using a dysprosium laser operating at $\lambda = 2.36$ μm. Based on the quadratic dependence of the intensity of the LA(709) line on the intensity of the exciton line over a wide density range (from 10^{12} to $5 \cdot 10^{14}$ cm^{-3}) it was concluded that the LA(709) line is due to the radiative annihilation of the electron and hole in a biexciton. Although this conclusion was not confirmed by later work, the experimental data concerning the kinetics of the free exciton line LA(714) and the LA(709) line contained in their work is of interest.

Hensel et al. [15, 65] have studied the kinetics of cyclotron resonance and microwave conductivity in Ge at 53 GHz. At low excitation levels ($\bar{n} < 10^{12}$ cm^{-3}) the resonance microwave conductivity signal is an exponential with a 0.8 μsec time constant; the authors assign it to a free exciton. Above threshold ($\bar{n} > 10^{12}$ cm^{-3}), the signal shows a second exponential with time constant $\tau = 39$ μsec at T = 1.14°K which results from the destruction of the drop because of Auger processes and radiative recombination. From these experimental data the volume $V_c = 1.6 \cdot 10^{-5}$ cm^3 and the particle density in the drop $n_0 = 1.3 \cdot 10^{17}$ cm^{-3} have been calculated.

The results of Hensel et al. were later generalized [66] to study drop kinetics using cyclotron resonance and measurements of the dielectric constant at microwave frequencies. Two temperature regimes were observed in cyclotron resonance studies when pulsed excitation was used. Below 2.5°K the cyclotron resonance kinetics are quasiexponential with a time constant of about 40 μsec; this corresponds to the extinction of a drop because of carrier recombination in the bulk of the drop. For temperatures above 2.5°K where the dominant decay mechanism for a drop is thermal evaporation of excitons from its surface, the cyclotron resonance signal loses its exponential character and decreases to zero in a time t_c which depends on the temperature. The temperature dependence of t_c enabled them to construct the liquid—gas phase diagram and to find the exciton work function to be $\Phi = 18 \pm 2$°K. The kinetics for changes in the dielectric function with time at T = 1.1°K are different from the kinetics for cyclotron resonance. First, the attenuation is much faster than 40 μsec; second, ε' changes sign and becomes negative when t > 110 μsec. The author explained this fact as follows: The presence of an electron-hole drop increases the actual positive dielectric constant of the sample, whereas the change in ε' is proportional to the drop volume. Free carriers make a negative contribution to ε'. As recombination in the drop progresses, the contribution from the free carriers becomes dominant and causes ε' to decrease and ultimately change sign.

Gladkov et al. [40, 41, 67] have studied the kinetics of cyclotron resonance at 10 GHz and the radiative recombination of the LA(709) line simultaneously with strong optical excitation. They found that the ratio of the characteristic times for these two processes is about 2. Therefore they conclude that the LA(709) line is related to radiative annihilation of the biexciton rather than radiation from the drop. However, we note that this result is easily explained in terms of the drop model. Following Hensel [15, 65], Gladkov et al. [40, 41, 67] assumed that Auger electrons are emitted from the entire volume of the drop so that the number of such electrons is proportional to that volume, whereas the kinetics are luminescence kinetics. But in reality, as has been shown [68], because of the high collision frequency for carriers in the drop the Auger electrons are emitted from a thin ($\sim 10^{-5}$ cm) surface layer and their concentration decreases with a time of $^3/_2 \tau_0$, where τ_0 is the luminescence time constant. Thus, within experimental error these results [40, 41, 67] do not contradict the drop model. Gladkov [41] has also measured the effective free-carrier collision frequency ($\leq 4 \cdot 10^9$ sec^{-1}) from the line width of the cyclotron resonance line when high levels of optical excitation are used.

It was assumed from the beginning of studies on electron-hole drops that the drops were electrically neutral. However, one would expect a drop to develop a charge because of differences in the binding energies of electrons and holes. The first experimental work to measure the charge of a drop was carried out by Pokrovskii and Svistunova [69]. They studied the surface distribution of recombination radiation from drops in Ge in the presence of a weak constant electric field. It was discovered that in fields of the order of 2.5 V/cm the radiation intensity distribution was always shifted toward the anode with respect to the center of the excitation region irrespective of the polarity of the applied field. They therefore concluded that the drops drift in the electric field because they have a net negative charge. They estimated both the drop charge ($\sim 100\,e$) and mobility [$\mu \geq 300$ cm^2/(V-sec)].

Many authors believe that, because the drops are negatively charged, they must electrostatically repel one another. Pokrovskii and Svistunova [70] studied the effect of uniaxial compression on the drop charge in Ge, and observed that the drops remain negatively charged up to pressures of 170 kg/cm^2 and then become positively charged. This is a result of changes in the work functions of electrons and holes in the condensed phase as deformation occurs.

An effective method for distinguishing between radiation from biexcitons and drops is to measure the polarization of the luminescence as the Ge sample is deformed. Bir and Pikus [71] have performed calculations to show that uniaxial stress splits the ground state of the exciton and biexciton such that, if the splitting is of the order kT, a significant fraction of the particles will be found in the lower level. Exciton and biexciton radiation should be polarized even at very low pressures, but the luminescence from the condensed phase will be polarized at very high pressures where the splitting of the valence band becomes comparable with the Fermi energies of the electrons and holes (i.e., energies much greater than kT). In the first experimental work along these lines Rogachev et al. [11] found that the radiation in the LA(709) became polarized at very low pressures, indicating that the line should be assigned to the biexciton. However, subsequent studies by Alekseev et al. [72] and still later by Pokrovskii and Svistunova [73] established that the degree of polarization in this line indicates that it originates from a drop. These investigations are continuing.

The work being performed by Jeffries' group is of interest. They have conducted a large number of studies which have revealed some new and interesting properties of the condensate and improved our knowledge concerning the parameters associated with the free-exciton—electron-hole drop system. Detailed studies of luminescence [74] have indicated hysteresis in the LA(709) line; at a given illumination power the intensity of this line is higher when the specified power is approached from the high excitation side than from the direction of lower excitation power. This hysteresis indicates that an exciton gas must be highly supersaturated in order to form drops.

Lo [75] has carried out spectroscopic studies on Ge to determine the condensation energy and electron-hole drop density. Earlier work [76] had indicated some broadening of the free-exciton line as compared with the theoretical value, leading to uncertainty in the calculated condensation energy for electrons and holes forming drops. Detailed measurements by Lo [75] at T = 2-4.2°K showed no differences between the experimental and theoretical line shapes; this allowed the condensation energy to be obtained: $\Phi = 2.06 \pm 0.15$ meV. The temperature dependence of the particle density in the drop was also obtained: $n_0(T) = 2.38 \cdot 10^{17}(1 - 0.01T^2)$ cm^{-3}. The critical temperature for drop formation was determined to be 8.3°K. In addition, the free-exciton line shape was used to obtain the splitting of the exciton level (0.7 meV).

The work by Jeffries' group on threshold phenomena in the electron-hole condensate is reviewed in [77]. It contains a discussion of a phenomenological model for the free-exciton—electron-hole drop system, including the process by which a drop is formed, which adequately describes the various features of the phase transition. It cites numerical data for the most important parameters and points out the good agreement between experiment and theory. Figure 1 shows the phase diagram for the transition from a free exciton gas to electron-hole drops in Ge [77]. The left side of the phase diagram bounds the region for the gas phase; these

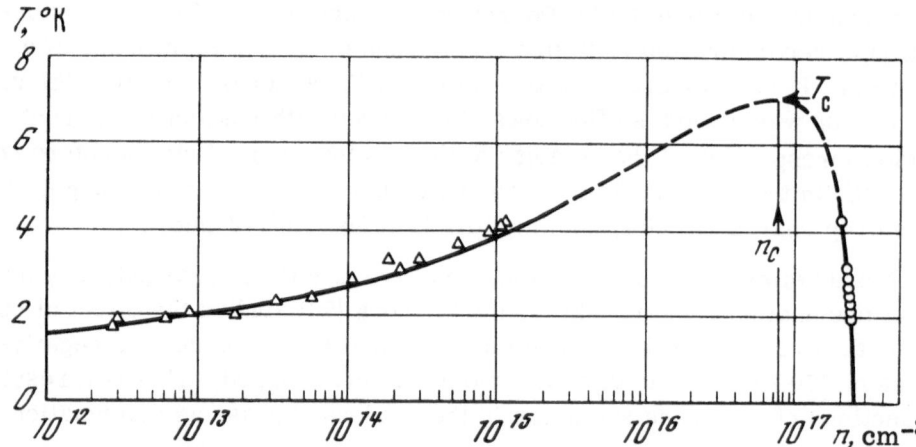

Fig. 1. Phase diagram for the phase transition from a free-exciton
gas to an electron-hole liquid in Ge [77]. T_c is the critical temperature and $n_c = 7.5 \cdot 10^{16}$ cm^{-3} is the critical density.

data come from studies of the threshold for the appearance of the LA(709) line as the temperature is lowered. The right side comes from spectroscopic measurements; the equilibrium
carrier concentration in a drop is determined from the LA(709) line width.

The foregoing discussion has dealt with the usual drop of about 10 μm radius and 40 μsec
lifetime. Recently there have been rather interesting reports of macroscopic (up to 1 mm
radius) drops. Markiewicz et al. [55] have observed large-radius drops (100 μm) with 500 μsec
lifetimes in ultrapure Ge compressed along the ⟨100⟩ axis. These results were obtained in
microwave-size resonance studies. The Ge crystal was placed in a special holder which ensured nonuniform compression, and introduced into a microwave cavity. An external magnetic
field of 3–20 kOe was applied to the sample parallel to the ⟨100⟩ axis. A series of new resonance lines related to absorption of microwave power by the drops was observed; the drop size
can be obtained from the magnitude of the resonant magnetic field. The microwave size resonance is interpreted as standing Alfvén waves in the drop. The lifetime of a large drop is obtained from the kinetics of the size resonance.

Macroscopic drops have been observed [78] through the spatial distribution of the LA(709)
luminescence line. The drops are formed in the regions with greatest pressure and have radii
varying from 30 to 300 μm depending on the intensity of the laser used for excitation. The recombination radiation kinetics agree with the kinetics of microwave size resonance. The
authors suggest that the drop lifetime increases because the equilibrium density of electron-
hole pairs in large-radius drops produced by high pressure is much lower (~10^{16} cm^{-3}) than
that of conventional drops.

Wolfe et al. [56] were the first to photograph a large-radius electron-hole drop. The
image was obtained by focusing the drop's recombination radiation ($\lambda = 1.75$ μm) onto a Vidicon
surface. The photographs confirm the data concerning large drops which have been obtained
from size resonance and luminescence.

Thus, a great body of experimental work has now confirmed a new phenomenon in semiconductor physics, the condensation of excitons into electron-hole drops. The problem now is
to study more completely the properties and parameters of the condensate.

2. MICROWAVE CONDUCTIVITY AND BREAKDOWN OF EXCITONS IN Ge

This review is devoted to a detailed examination of the microwave conductivity of pure
Ge at helium temperatures and with laser excitation, for the purpose of studying the exciton

condensation process. The conductivity in both weak and strong microwave fields using optical excitation is examined to reveal the kinetics. In this section we shall describe the experimental methods developed by us for observing the microwave conductivity and breakdown effect.

2.1. Apparatus for Studying Microwave Conductivity, Cyclotron Resonance, and Breakdown Effect at 10 GHz and 1.3-4.2°K in Ge

The apparatus diagrammed in Fig. 2 has been constructed to study the microwave conductivity of Ge using optical excitation. It consists of a 10-GHz microwave spectroscope with reflecting cavity and a pulsed Nd-YAG 1.06 μm laser. The microwave spectroscope was developed in the Vibrations Laboratory of the Physics Institute, Academy of Sciences, for video and superheterodyne microwave spectroscopy [79]. The design has been used to study electron resonance spectra and relaxation processes in paramagnetic cyrstals. Such microwave spectroscopes are readily adapted to study of the kinetics of microwave conductivity, the exciton breakdown effect in both cw and pulsed microwave fields, and cyclotron resonance. Since these instruments have been described in detail elsewhere [79, 80] we shall discuss just the principle of operation and their most important features relative to our present problem.

The microwave power from the signal generator (klystron) is introduced to the reflecting cavity containing the sample by means of a ferrite circulator. The reflected power goes to either a video or superheterodyne receiver. In the latter case, after the signal is converted

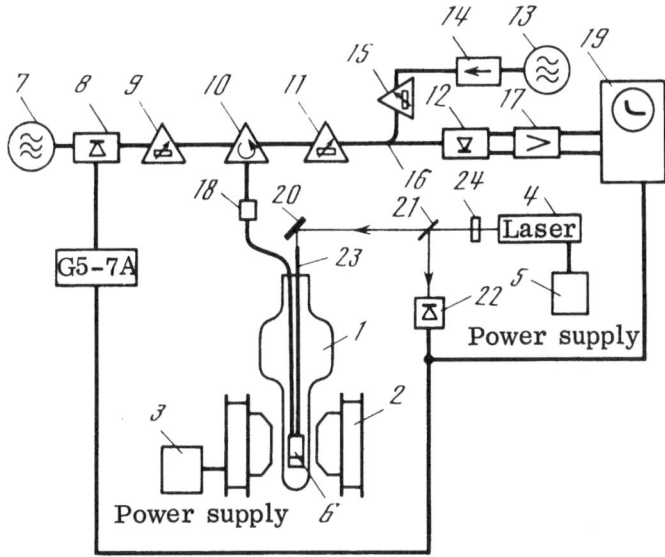

Fig. 2. Block diagram of the apparatus for measuring microwave conductivity and breakup of excitons in Ge using pulsed laser excitation (T = 1.3-4.2°K). 1, cryostat; 2, electromagnet; 3, magnet power supply; 4, Nd-YAG laser; 5, laser power supply; 6, cavity with sample; 7, signal klystron; 8, microwave modulator; 9, 11, 15, attenuators; 10, ferrite Y-circulator; 12, diode mixer; 13, heterodyne klystron; 14, ferrite rectifier; 16, directional coupler; 17, IF amplifier with second detector; 18, impedance transformer; 19, oscillator; 20, mirror; 21, splitter; 22, photodiode; 23, light pipe; 24, neutral filter.

in the mixer and amplified at an intermediate frequency, it is detected by a second detector, further amplified by a wide-band video amplifier, and displayed on the oscilloscope screen.

The mixer of the superheterodyne spectroscope uses a balance circuit to eliminate noise from the heterodyne. Crystals such as D-403 diodes are used as mixers. The IF amplifier center frequency is 35 MHz and has a π-shaped pass band of about 10 MHz.

An important element in the spectroscope is the microwave power modulator which is found in the signal klystron circuit. Power modulation was used to study the breakdown effect in the pulsed mode. Our spectroscope used a modulator based on semiconductor diodes. The operation of such modulators depends on changes in diode resistance as a bias voltage is applied. The matching between the diode and waveguide changes as the diode resistance changes. This allows one to modulate the power entering the waveguide from the diode. We used a single modulator section with a D-13 germanium diode, permitting about 20 dB reduction in power. The semiconductor modulator has good time characteristics: the switching time is less than 1 μsec. The modulator provided square power pulses with leading edges about $\Delta\tau \simeq 0.2$ μsec wide. The modulating voltage goes from the G5-7A pulse generator to the diode.

The superheterodyne microwave spectroscope was used to study microwave conductivity and cyclotron resonance with small levels of optical excitation when high receiver circuit sensitivity is required. A videospectroscope was used in the studies of breakdown effects. Here the receiver circuit consisted of a microwave detector and an amplifier.

The most important characteristic of the microwave spectroscope used in the study of microwave conductivity kinetics and in determining the recombination times for carriers, excitons, and electron-hole drops is the resolution time. It is determined by two factors, the time constant $\Delta t = Q/\omega_r$ for establishing oscillations in the microwave cavity (Q is the cavity Q-factor and ω_r is the cavity resonant frequency), and the pass band of the receiver circuit. For the rectangular cavity which we used, in which H_{102} oscillations were excited, the Q at helium temperatures was about 10^3 (this is voltage Q with sample losses taken into account); thus Δt was about 10^{-7} sec. The pass band of the receiver circuit is determined by the IF amplifier frequency band ($\Delta f = 10$ MHz) when the microwave spectroscope is operated in the superheterodyne mode, and by the low-frequency amplifier band which was the oscillosope amplifier ($\Delta f = 25$ MHz) or a transistor amplifier with a pass band of $\Delta f = 10$ MHz. Therefore, the resolution of our microwave spectroscope was limited basically by the cavity time constant and was $\Delta t_{m.s} \simeq 10^{-7}$ sec.

The sensitivity of the microwave spectroscope in the superheterodyne mode was about 10^9 free carriers per cubic centimeter as measured from the lowest recorded signal in microwave conductivity measurements after a laser pulse of known power at room temperature. Since the cavity Q at room temperature was about a factor of 2 lower than the Q at helium temperatures, the corresponding correction has been made to the calculation.

The sample was excited optically by means of a neodymium-YAG laser operated in the Q-switched mode. We shall discuss the laser briefly. The active element is a garnet crystal containing Nd^{3+} (0.8 at.%) in the shape of a cylinder 80 mm long by 6 mm diameter. The pumping lamp was a cw krypton lamp located in a silvered-quartz illuminator. In order to cut off the ultraviolet radiation from the lamp it was placed inside a tube of yellow glass. The maximum current through the lamp was 30 A at 200 V. The multilayer dielectric mirrors on glass substrates had the following characteristics: front mirror reflectivity 93%, back mirror 98%. The laser power supply circuit is shown in Fig. 3. The capacitors are charged up to 1000 V to fire the flash lamp and to power the flash lamp in the steady-state mode.

This laser can be operated either in a cw mode or in the Q-switched mode. The average cw power is about 0.5 W. The laser can generate light pulses 100 nsec long with a maximum

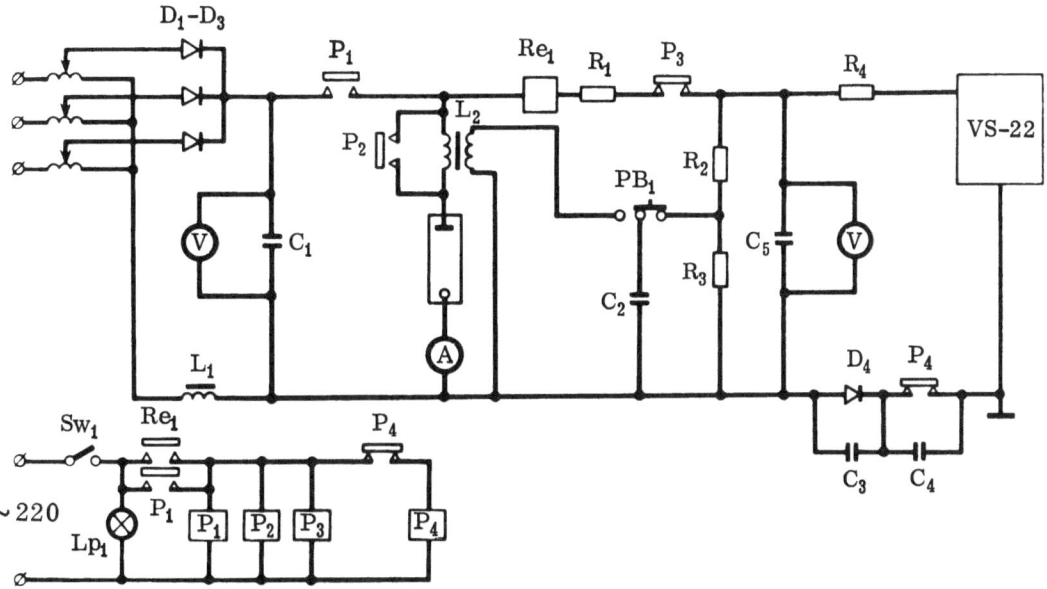

Fig. 3. Block diagram of the power supply for the Nd-YAG laser. P_1-P_4 are ac relays; R_1-R_4 are resistors; C_1-C_5 are capacitors; D_1-D_4 are diodes; L_1, L_2 are inductors; Re_1, RE_2 are relays; V is a voltmeter; A is an ammeter; Sw_1 is a switch; Lp_1 is a lamp; PB_1 is a pushbutton; VS-22 is a high-voltage electron-stabilized rectifier.

energy of 10^{-4} J/pulse at a repetition rate of about 100 Hz. The Q is switched by rotating a mirror.

The laser radiation is attenuated by calibrated neutral density filters. The laser pulse is split and falls simultaneously on both the sample and a photodiode whose signal triggers the G5-7A pulse generator which is connected to both the microwave modulator and the oscilloscope sweep. When recording cyclotron resonance spectra of Ge excited by cw laser radiation we used an electromagnet whose field strength could be as high as 10 kG. The magnet was powered by a regulated supply.

Two types of helium cryostats were used. One is a metal KR-21 (FIAN) cryostat* with a narrow shaft to fit between the magnet pole faces, and the second has optical windows but is based on the KR-21 design. The quartz windows in the helium section were glued to the Invar mandrels with PR epoxide, a glue developed by the Cryogenics Division of the Physics Institute, Academy of Sciences of the USSR [81].

A VN-4 pump was used to pump the helium vapor. The lowest temperature reached in these experiments was about 1.3°K. The Ge samples were located inside the microwave cavity. When the optical cryostat was used the sample was put into a quartz tube opposite an opening in the narrow wall of the cavity. When the KR-21 cryostat was used the samples were attached to a light pipe and could be positioned at various points in the cavity.

2.2. Microwave Conductivity and Breakdown

Experiments

The interaction of electromagnetic radiation with free carriers can be measured from changes in the complex dielectric constant whose real and imaginary parts are given by the

---

* The KR-21 cryostat was built from plans supplied by the Cryogenics Division of the Physics Institute, Academy of Sciences of the USSR.

well-known expressions [82]

$$\varepsilon' = -\frac{4\pi n_e e^2}{m\,(\omega^2 + \nu^2)}, \qquad \varepsilon'' = \frac{4\pi n_e e^2 \nu}{m\omega\,(\omega^2 + \nu^2)}, \tag{1}$$

where n_e is the density of free carriers, ω is the frequency of the electromagnetic field, ν is the effective collision frequency, and e and m are the carrier charge and mass.

A microwave cavity can be used to record both components of the complex dielectric function $\varepsilon = \varepsilon' + i\varepsilon''$ of a sample; changes in ε' cause the cavity to be retuned, and changes in ε'' are related to absorption of microwave power in the sample which alters the cavity Q. The sign of $\Delta\varepsilon'$ can be determined from changes in the resonant frequency.

Optical excitation of the sample produces free carriers which, after a very short time ($\ll 1\ \mu$sec), decay into excitons. The excitons subsequently condense into electron-hole droplets. Studies have shown (see Section 3) that at 10 GHz the changes in microwave conductivity, or the equivalent change in the dielectric constant, under optical excitation are due mainly to the free carriers. Thus, the signal observed by a microwave spectroscope as a change in the power reflected from the cavity during optical excitation is, in general, a function of ε' and ε'' for free carriers. The shape of the signal describes the kinetics for the number of free carriers available after a pulse of optical radiation, reflecting the carrier recombination processes and their connection with excitons and electron-hole droplets (EHD). Thus, although one directly records changes in sample conductivity in these experiments, the signal also contains direct information about excitons and drops since all three components of the system (free carriers, excitons, and EHD) are in equilibrium with one another.

Experimentally, for T ≤ 3.0°K, the signal reflected from the cavity is the combination of two exponentials with time constants τ_1 and τ_2. As the temperature is lowered to 2°K, τ_2 increases and then begins to vary. It is natural to assume that the first, shorter time constant accounts for the reduction of the laser-generated carriers into excitons and drops while the longer time constant τ_2 describes a new exciton phase.

In order to follow the shape of the signal as a function of the laser pulse energy we studied the microwave conductivity of Ge at different excitation levels beginning with very low levels. We used the superheterodyne spectroscope which enabled us to record free carriers with densities $\gtrsim 10^9$ cm^{-3}.

Nonlinear distortions of the signals can appear in studies of the kinetics of microwave conductivity. These nonlinearities can be caused by nonlinear response in the cavity microwave spectroscope [80] and in the detector. Such distortions can have a significant effect on the accuracy of time constant measurements and measurements of the amplitudes which describe the binding and recombination processes of the carriers. In addition, nonlinearities in the microwave spectroscope can significantly affect estimates of the free carrier concentrations resulting from the destruction of excitons by the microwave field. Thus, we have analyzed in detail the various sources of nonlinearity which might be operating in our microwave spectroscopes by using the well-known theories for EPR microwave spectroscopes [80].

The most important source of nonlinearity in a microwave spectroscope is the "resonator effect," a generally nonlinear relationship between the cavity power and ε' and ε''. The condition for linearity in the microwave spectroscope is satisfied only at low carrier concentrations, just as in EPR [80]. When the exciting laser pulse has a high intensity the starting density of nonequilibrium carriers in the sample can reach 10^{16}-10^{17} cm^{-3}. The values of ε' and ε'' corresponding to such densities can result in marked detuning of the cavity, which the amplifier cannot handle, and this significantly changes the Q so that the shape of the signal becomes a very nonlinear function of both ε' and ε''. In addition to the apparatus distortions at high free

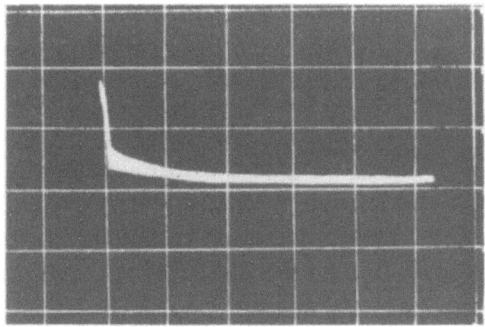

Fig. 4. Signal from microwave conductivity of Ge using pulsed laser excitation with T = 1.3°K and $\bar{n} = 10^{14}$ cm^{-3}. The sweep is initiated at the end of the laser pulse. Abscissa 20 μsec/division, ordinate uncalibrated.

carrier concentrations, the signal shape is affected by field screening in the sample because of an increase in the modulus of the complex dielectric constant [82].

Figure 4 shows that after 1-2 μsec the number of free carriers in the sample drops rapidly, the signal amplitude decreases, and the conditions for a linear relationship between p_r and ε', ε'' are satisfied. We conclude that the time constant for the fast exponential cannot be used to obtain exact characteristics for the process of turning carriers into excitons and EHD because this exponential is very distorted, especially at the beginning. However, the slower exponential with time constant τ_2 correctly reflects the drop recombination process because the linearity conditions are now satisfied.

The pass band of the recording apparatus was wide enough for our experiments. The time resolution of the equipment was limited by the cavity time constant, as we have discussed earlier (see Section 2.1); it is about 10^{-7} sec long.

The breakdown effect, which is the destruction of excitons by the microwave field, was observed when microwave power greater than 5 mW is introduced into the cavity. To study this effect we used modulated microwaves in which the recorded signal was obtained from a low level of leaking power (P < 0.05 mW) while breakdown was accomplished by more powerful pulses of microwaves. The amplitude, duration, and delay of these pulses relative to the laser pulses could be controlled over wide limits.

The temporal characteristics of the diode microwave modulator (the leading edge of the microwave pulses from our spectroscope is about 0.2 μsec) had no real effect on the accuracy of the measurements of the breakdown characteristics (threshold, time development and decay). This is indicated by the clear features of the breakdown effect wherein the amplitude of the breakdown peak increased rapidly when the power of the microwave pulse just exceeded the threshold level.

In order to study the breakdown effect as a function of the relative strengths of the electric and magnetic components of the microwave field, the sample was located in different positions within the cavity, and different orientations of the sample relative to the directions of E and H were used.

In our experiments with the temperature dependence of the microwave conductivity and breakdown effect in the interval from 1.3-4.2°K the temperature was stabilized by a pump regulator which maintained the pressure over the liquid helium to within 1 mm Hg, thereby regulating the temperature to within about 0.01°K. The temperature was measured by means of a mercury manometer which measured the pressure of the saturated helium vapor.

2.3. Sample Characteristics

The sample was pure Ge because, as we have already discussed, Ge is a very suitable material for verifying theoretical models of exciton condensation. Most important is the fact that the band structure of Ge results in free-carrier lifetimes which are quite long (10^{-5} sec) at helium temperatures. This greatly simplifies studies of the time and concentration dependences of processes taking place in the crystal after laser excitation.

In many respects an exciton is much like a hydrogenic impurity. In order to study collective effects in a system of excitons it is important to use samples with a minimum of impurities. This is especially important when studying breakdown effects because hydrogenic impurities break down in electric fields of about the same strength as those used to break apart excitons. It was found necessary to study many samples with different densities of residual impurities in order to isolate the exciton breakdown effect from the impurity breakdown and in order to obtain reproducible results with different samples.

We studied five groups of Ge samples having residual impurity concentrations $N_A + N_D$ ranging from $5 \cdot 10^{13}$ to 10^{11} cm^{-3}. In addition the exciton breakdown effect in pulsed microwave fields was studied using ultrapure Ge samples with $N_A + N_D \approx 10^{10}$ cm^{-3} obtained from General Electric [83]. Typical sample dimensions were $5 \times 5 \times 0.5$ mm; we did not use samples thinner than 0.3 mm. The samples were polished and etched in boiling H_2O_2 + NaOH in order to reduce surface recombination. Some of the samples had smooth polished surfaces.

The starting density of free carriers generated by the laser pulse was determined from the light intensity falling on the sample, and was assumed to be uniformly distributed throughout the sample. The latter assumption was justified by the observation that, while the carriers are generated in a thin layer near the surface (the optical absorption coefficient of Ge at 1.06 μm is 10^4 cm^{-1}), they are diffused uniformly to a depth of ≤ 1 mm at helium temperatures [84] in about 10^{-6} sec.*

3. MICROWAVE CONDUCTIVITY AND BREAKDOWN OF EXCITIONS IN Ge: RESULTS

As discussed in the Introduction and Section 1, the electrical properties of Ge at 10-100 GHz under strong laser excitation are of great interest for studying the collective properties of excitons at high densities (in particular when treating exciton condensation). A number of studies [15, 40, 41, 42, 65, 66] have been devoted to the kinetics of microwave conductivity and cyclotron resonance in Ge using laser excitation.

At the same time that the studies just mentioned were being carried out we began our microwave experiments on Ge and observed [18, 20] a new effect, the breakdown of excitons in the microwave field. The detailed study of this effect provided new data on the properties of excitons, data which were very informative concerning the collective properties of excitons. Our experiments used both cw and pulsed microwave fields. In this section we will describe the results of the microwave conductivity experiments in Ge and the breakdown effect, and we present a theoretical interpretation of the results.

3.1. Microwave Conductivity and Breakdown

in a Continuous Field

These experiments were performed with the 10-GHz superheterodyne microwave spectroscope (see Fig. 2). The samples were Ge with residual impurity concentrations ranging from $5 \cdot 10^{12}$ to 10^{11} cm^{-3}. Typical sample dimensions were $2 \times 2 \times 0.5$ mm.

Figure 5 shows typical microwave conductivity signals observed using pulsed laser excitation and low microwave power levels (P < 0.05 mW) with T = 4.2°K. The signal changes

---

* The density of nonequilibrium carriers can also be determined from the kinetics of microwave absorption [85]. This method is described in detail in another article of this volume (see p. 45).

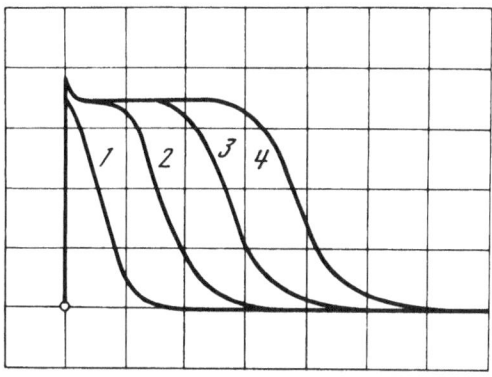

Fig. 5. The microwave conductivity signal for Ge using pulsed excitation with different laser power levels (T = 4.2°K). 1) $\bar{n} = 10^{13}$ cm^{-3}; 2) $\bar{n} = 5 \cdot 10^{13}$ cm^{-3}; 3) $\bar{n} = 10^{14}$ cm^{-3}; 4) $\bar{n} = 5 \cdot 10^{14}$ cm^{-3}. The sweep is initiated by the trailing edge of the laser pulse. Abscissa scale is 10 μsec/division, the ordinate is uncalibrated.

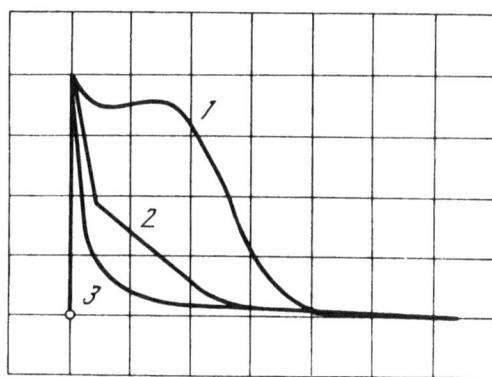

Fig. 6. Microwave conductivity signal in Ge at various temperatures. 1) T= 4.2°K; 2) T = 2.8°K; 3) T = 1.3°K. Starting density of electron-hole pairs is n = 10^{15} cm^{-3}. The sweep is initiated by the end of the exciting pulse. Abscissa scale is 20 μsec/division, the ordinate is uncalibrated.

shape as the temperature is lowered. For T ≤ 3°K the microwave conductivity signal is the sum of two exponentials with different amplitudes A_1 and A_2 and different time constants τ_1 and τ_2. Temperature variations in the signal for a fixed level of laser pulse are shown in Fig. 6. It is observed that the fast exponential changes little as the temperature drops, but the slower term becomes elongated and its amplitude decreases.

A detailed study of the microwave conductivity of Ge following a laser pulse was conducted at T = 1.3°K with the sample located in the electric field of the cavity. Study of the signal shape in the presence of a constant magnetic field up to 10 kG showed that the amplitudes of the exponentials change in agreement with the cyclotron resonance lines. We therefore conclude that the observed signal is due to free carriers contributing to both the real and imaginary parts of the sample's dielectric constant. These contributions are given by Eqs. (1) (see Section 2.2).

Observation shows that the initial stage of the signal (the exponential with time constant τ_1) includes both absorption of microwave power and retuning, in agreement with Eq. (1) for relatively large carrier concentrations, where possibly $\nu \gtrsim \omega$. This initial phase evidently describes the process of binding carriers into excitons and EHD. The exponential with time constant τ_2 is primarily responsible for retuning the cavity through changes in ε', in agreement with Eq. (1) for small carrier concentrations and with $\nu \ll \omega$. We assume that the slower exponential is connected with the residual carriers which are in equilibrium with the excitons and EHD. Measurements of the sign of the frequency shift in the cavity, which is available from the slower exponential signal (τ_2), show that $\Delta\varepsilon'$ is negative. This confirms that the observed signal is related to residual carriers rather than excitons or EHD.

Similar experiments have been performed at 53 GHz by Hensel and Phillips [15]. They found that the change $\Delta\varepsilon'$ derived from the slow exponential is positive. This discrepancy might be explained by Eq. (1) in that at high frequencies the contribution to $B'\varepsilon'$ from free carriers is much lower. It is possible that the slower exponential signal seen by Hensel and Phillips was not due to free carriers.

By using a superheterodyne, high-sensitivity microwave spectroscope we carried out detailed studies of the changes in the microwave conductivity signal as the level of optical excitation is lowered. We observed that the slow exponential persisted down to excitation levels corresponding to initial concentrations of electron-hole pairs of the order of $\bar{n} \simeq 10^{13}$ cm^{-3}. As $\bar{n}$ varied from 10^{16} to 10^{13} cm^{-3} the amplitude decreased but the time constant τ_2 remained unchanged. When $\bar{n} < 10^{13}$ cm^{-3} the signal becomes comparable to the noise and we cannot draw any conclusions concerning a possible threshold for the appearance of EHD which are responsible for the slow exponential in the signal.

The above experiments on the microwave conductivity of Ge were performed using a cw power input of less than 0.05 mW to the cavity. As the power was raised above 5 mW we observed a sharp increase in conductivity (the breakdown peak) with a leading edge of 0.6 μsec and a trailing edge of 2-3 μsec. This peak is delayed relative to the laser pulse by some period t_d which depends on the exciting intensity and the microwave power. A typical value of t_d is 100 μsec for $\bar{n} \simeq 10^{15}$ cm^{-3} and P = 10 mW.

Figure 7 shows the microwave conductivity signals obtained with different microwave power levels when the sample is placed at the position of maximum electric field with T = 1.3°K. For P $\simeq$ 0.5 mW one observes a bump at the end of the slow exponential (Fig. 7b, c) which shifts towards the left as the microwave power increases. When P is greater than 5 mW it changes into a sharp peak which we assign to the destruction (or breakdown) of the excitons. Figure 8 shows an oscillogram corresponding to this signal.

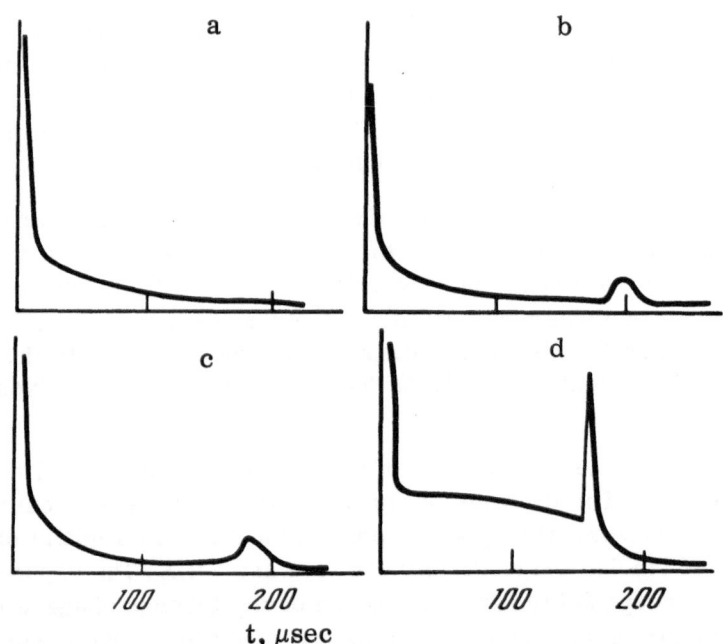

Fig. 7. Microwave conductivity signals obtained with different microwave power levels with the Ge sample located at a maximum of the field. T = 1.3°K. a) P < 50 μW, E < 1 V/cm; b) P ~ 0.5 mW, E ~ 5 V/cm; c) P ~ 1 mW, E ~ 7 V/cm; d) P ~ 5 mW, E ~ 15 V/cm. Ordinate is in relative units. Sweep starts at the end of the exciting pulse. $E_p = 10^{-6}$ J.

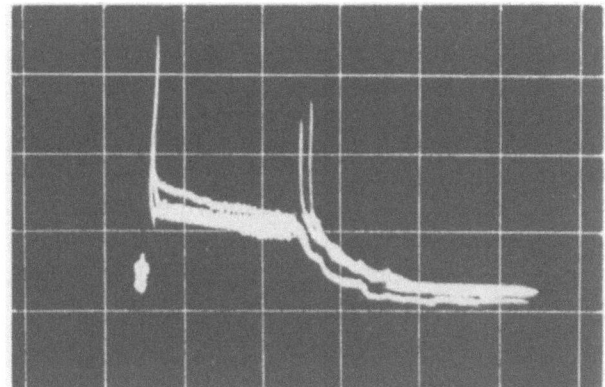

Fig. 8. Oscillogram of the microwave conductivity signal corresponding to case d in
Fig. 7. Abscissa scale is 50 μsec/division.

Further examination of the breakdown effect shows that it is observed when the pump energy of the radiation is 10^{-7}-10^{-6} J per pulse (corresponding to a starting nonequilibrium density of carriers of the order of ~10^{15}-10^{16} cm^{-3}) and the temperature is 1.3-2.5°K. The delay of the breakdown pulse relative to the excitation depends on the excitation level: The lower the excitation level the sooner breakdown appears. Delay times ranging from 30 to 160 μsec have been observed. In addition, the time at which the breakdown peak appears depends on the microwave power level. When P = 5 mW the delay time is about 160 μsec, but when the power is increased to 50 mW the delay drops to 25 μsec. Figure 9 shows the dependence of the breakdown delay time on temperature for fixed excitation and microwave power levels.

An external magnetic field has a significant influence on the breakdown process. For fields corresponding to the cyclotron resonance lines in Ge the delay time t_d decreases, and when H > 2 kG breakdown disappears. When the sample was placed at a maximum of the micro-wave magnetic field in the cavity an absorption signal was observed which was 100 times smaller than that seen when the sample was located in the microwave electric field. Furthermore, the microwave conductivity signal observed under these conditions for any microwave power level and any laser power level was practically identical with that shown in Fig. 7a; i.e., breakdown is not present.

Below we will discuss some possible models for exciton breakdown based on the biexciton gas and electron-hole droplets. But first we must present results of experiments on the

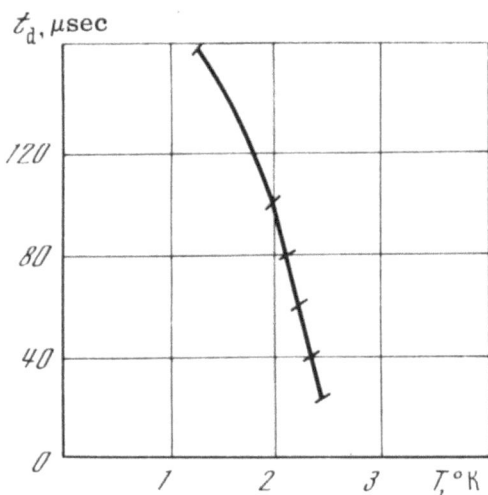

Fig. 9. Temperature dependence of the breakdown delay time for a fixed microwave power
P = 10 mW and a laser excitation level producing $\bar{n} = 10^{15}$ cm^{-3}.

breakdown effect in pulsed microwave fields. Such studies provide much data on the nature of the breakdown, data which are very important in selecting a mechanism for exciton destruction and modeling the collective state of excitons at moderate densities.

3.2. Exciton Breakdown in Pulsed Microwave Fields

Assuming that the breakdown process for excitons is similar to that operating in laser and microwave breakdown of gases and solids [86-88], we have investigated microwave breakdown in Ge in the pulsed mode [19, 20] by varying the pulse lengths and the delay between the microwave pulses and the exciting laser pulse.

The experiments were performed using the apparatus shown in Fig. 2 and discussed in Section 2.1. We used Ge samples with residual impurity concentrations $N_A + N_D$ varying between $5 \cdot 10^{12}$ and 10^{10} cm^{-3}; the samples were plates with typical dimensions of $5 \times 5 \times 0.5$ mm.

It was observed that when a microwave pulse was delayed a time t_d relative to the laser pulse, a sharp burst of conductivity was measured when the microwave power exceeded the threshold value P_d (Fig. 10). The breakdown has a very clear threshold character; the peak rapidly increases even for small increases in power above threshold $(P - P_d)/P_d \sim 0.01$. This enables one to accurately determine the breakdown threshold. Since the threshold power associated with breakdown is one of the most important characteristics, we measured the threshold under various conditions. By threshold we mean the smallest amount of microwave power introduced into the cavity which will produce breakdown.

It was assumed that the breakdown threshold can depend on the relative magnitudes of the electric and magnetic fields of the microwave field, on the dimensions of the exciton complexes, and on the initial concentrations of carriers and excitons. Therefore it was necessary to study the dependence of P_d on the sample position within the cavity, on the duration τ and delay t_d of the microwave pulse relative to the laser pulse, and on the intensity of the laser pulse. We shall discuss below the experimental results obtained for each of these dependences.

A. Dependence of P_d on the Sample Location within the Cavity

In a rectangular microwave cavity the E and H fields are quite well separated, so that with samples as thin as those we used (≤ 1 mm) it was possible to situate them in either an electric or a magnetic field to determine which is responsible for breakdown. Figure 11 shows the dependence of the breakdown threshold on the sample position along the z axis of the cavity

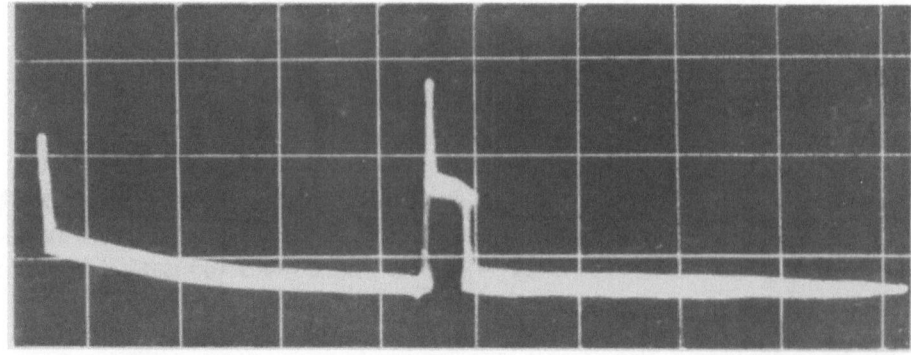

Fig. 10. Oscillogram of the signal reflected from the cavity after the Ge has been excited by a laser pulse. The breakdown peak is visible at the peak of the microwave pulse which is delayed about 80 μsec relative to the laser pulse. Scale: 20 μsec/division.

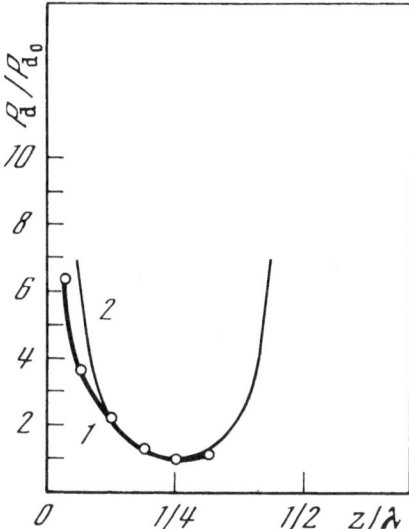

Fig. 11. Breakdown threshold P_d as a function of sample position along the z axis of the cavity (curve 1). For comparison we show the distribution of the component $(E_{y_0}/E_y)^2$ of the microwave electric field in the cavity (curve 2). P_{d_0} is the breakdown threshold at the maximum of the electric field.

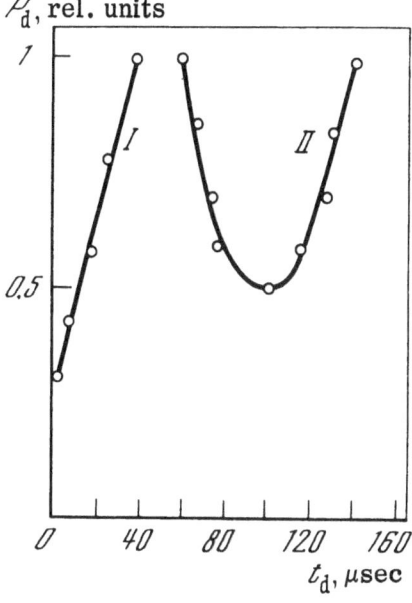

Fig. 12. Dependence of the breakdown threshold on the time delay between the microwave pulse and the laser pulse. The microwave pulse is 0.2 μsec long and the initial density of electron-hole pairs is $\bar{n} = 8 \cdot 10^{14}$ cm^{-3}.

for a fixed microwave pulse length and delay time and a specified laser power. We see that the threshold is a minimum when the sample is in the electric field and that there is no breakdown when it is in the microwave magnetic field. The shape of the P_d (z) curve departs somewhat from the function E_y^2 (z) = $E_0^2 \sin^2 k_z z$ for the y-component of the electric field for the H_{102} oscillations of an empty cavity. This discrepancy can be attributed to a redistribution of the field in the cavity because of the retuning induced by the presence of the sample.

B. Dependence of P_d on the Microwave Pulse Delay Time t_d

Figure 12 shows the dependence of the breakdown threshold on the delay between the laser pulse and microwave pulse of fixed duration ($\tau \simeq 0.2$ μsec) for an initial density of nonequilibrium carriers of $n \simeq 8 \cdot 10^{14}$ cm^{-3}. The curve clearly displays two branches, the first (I) for times between 3-40 μsec and the second (II) between 50 and 150 μsec with a minimum near 100 μsec.

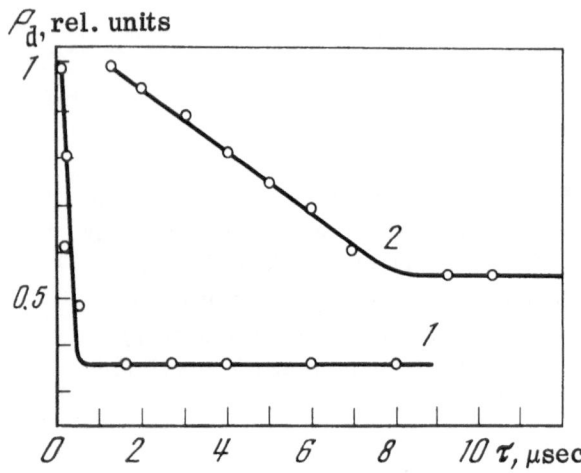

Fig. 13. Dependence of the breakdown threshold on the duration of the microwave pulse. 1) for exciton breakdown; 2) for impurity breakdown. The exciton breakdown data were obtained using a laser pulse−microwave pulse delay time of $t_d = 100$ μsec and $\bar{n} = 10^{15}$ cm^{-3}. Impurity density was ∼$5 \cdot 10^{12}$ cm^{-3}.

Breakdown has a somewhat different character on the two branches. The difference is due to both the sharpness of the breakdown threshold and to its time development. In region I the existence of a threshold is not so clearly evident and the amplitude of the breakdown peak increases rather slowly with the power in the microwave pulse. On this branch the leading and trailing edges are $\tau' \simeq \tau'' \simeq 0.2$ μsec. On branch II the breakdown threshold is very clearly expressed and the edges of the breakdown peak are $\tau' < 0.1$ μsec and $\tau'' < 0.2$ μsec.

Note that the first branch of $P_d(t_d)$ exists only for rather high levels of laser power. When the excitation intensity is such that the initial carrier density within the sample is $\bar{n} \le 2.5 \cdot 10^{14}$ cm^{-3} the curve of $P_d(t_d)$ consists of just the second branch with the minimum.

We found that this breakdown effect occurred in all our samples regardless of residual impurity concentration; this included the ultrapure germanium sample ($N_A + N_D \le 10^{10}$ cm^{-3}). All the samples had the same characteristics (i.e., the same values of P_d, τ', and τ'').

C. Dependence of P_d on the Duration τ of the Microwave Pulse

Curve 1 of Fig. 13 shows the dependence of the threshold breakdown power on the duration of the microwave pulse when the initial density of carriers in the sample is $\bar{n} = 10^{15}$ cm^{-3}

Fig. 14. Dependence of the exciton breakdown threshold P_d on the microwave pulse delay time for various levels of optical excitation (initial density of electron hole pairs). 1) $\bar{n} = 10^{15}$ cm^{-3}; 2) $\bar{n} = 5 \cdot 10^{14}$ cm^{-3}; 3) $\bar{n} = 2.5 \cdot 10^{14}$ cm^{-3}. The microwave pulse was 0.2 μsec long.

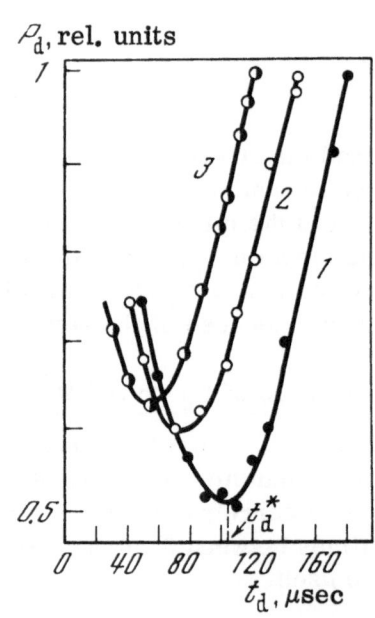

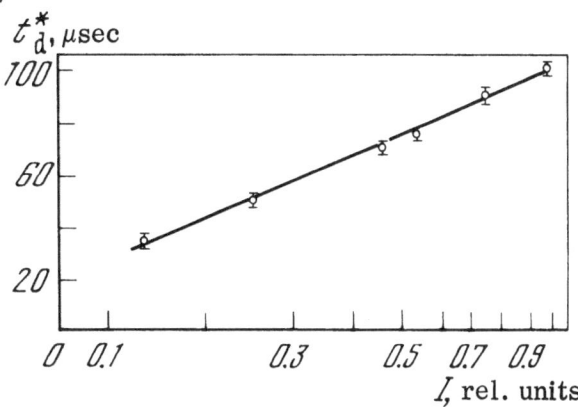

Fig. 15. Dependence of the delay time at the minimum exciton breakdown threshold t_d^* on the laser intensity. The abscissa is logarithmic. The point I = 1 corresponds to an initial density of $\bar{n} = 10^{15}$ cm^{-3} pairs and the slope of the straight line is the time constant $\tau_0 = 35$ μsec.

and the delay between the laser and microwave pulses is 100 μsec. The curve decreases as the duration increases and finally becomes a flat function when $\tau = 0.5$ μsec.

Experiments with breakdown due to two successive microwave pulses of equal duration and amplitude were also conducted. It was observed that if the time interval between these two pulses was shorter than 10-15 μsec, breakdown was achieved only by the first pulse; if the separation of the pulses is longer than 10-15 μsec the second pulse will also cause breakdown.

D. P_d as a Function of the Intensity of Optical Excitation

We have studied in detail the breakdown threshold dependence on the initial density of electron-hole pairs as determined from the intensity of the laser pulse. Figure 14 shows a family of curves for various levels of optical excitation intensity I. We see that $P_{d\,min}$ shifts toward shorter delay times as the light intensity decreases. Figure 15 shows the dependence of $P_{d\,min}$ as a function of the intensity of the laser pulse.

We shall discuss in detail the experimental data presented above (see Section 3.4) using the model of breakdown in an exciton gas in the presence of electron-hole drops. For the moment, however, we shall confine ourselves to a few qualitative remarks. The dependence of the breakdown threshold on the sample's location within the cavity (Fig. 11) indicates at once that we are dealing with electrical rather than magnetic breakdown. Since the characteristics of this breakdown effect are the same regardless of the nature and concentration of impurities in all the germanium samples, it is clear that the effect is related to destruction of excitons or some sort of exciton complex by the microwave electric field, and is not connected with impurity breakdown (Section 3.3 deals with impurity breakdown). At the same time it is clear that the electric field of the microwave cavity cannot destroy electron-hole drops because one expects them to be heated more rapidly in a magnetic field. However, the dependence of the breakdown threshold on the delay between the microwave and laser pulses, when taken together with the other breakdown characteristics (duration of breakdown peak, breakdown by two pulses, etc.), indicates that the presence of EHD determines how microwave breakdown develops. A careful analysis of all the experimental data leads us to conclude that this effect is the electrical breakdown of a gas of free excitons in equilibrium with drops of an electron-hole plasma which, in turn, act as trapping centers for free carriers and therefore have a significant impact on the breakdown characteristics. The theoretical treatment of this model, presented below in Section 3.4, completely explains all the observed features of the breakdown.

3.3. Microwave Breakdown of Impurities

We have observed the microwave breakdown of impurities in addition to the breakdown of excitons. Impurity breakdown was observed only after the sample had been excited optically.

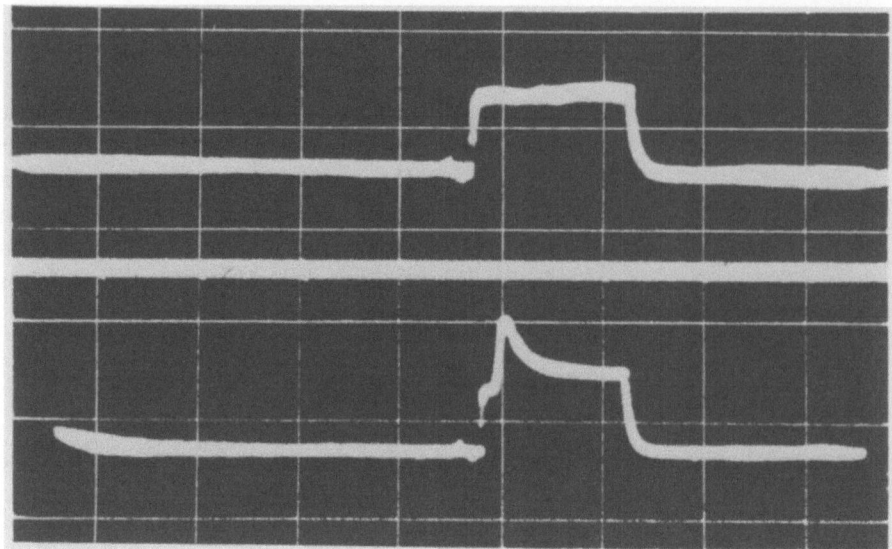

Fig. 16. Oscilloscope traces of microwave conductivity. The upper trace was obtained in the absence of impurity breakdown, the lower in the presence of impurity breakdown. The Ge sample had $N_A + N_D = 5 \cdot 10^{11}$ cm^{-3} and was held at T = 1.3°K. There was a 200 μsec delay between the microwave and laser pulses. Scale 50 μsec/division.

The characteristics of impurity breakdown are substantially different from those of exciton breakdown and change from sample to sample depending on the impurity concentration.

Figure 16b shows an oscilloscope trace of the impurity breakdown signal for a sample with $N_A + N_D = 5(10)^{11}$ cm^{-3} at T = 1.3°K. The leading edge of the breakdown signal is about 1 μm long and the trailing edge is ~10 μsec. The main difference between impurity breakdown and exciton breakdown is that impurity breakdown is independent of the time delay between the microwave and laser pulses. In samples with residual impurity densities of $5 \cdot 10^{11}$–$5 \cdot 10^{12}$ cm^{-3} breakdown will occur as much as 10 μsec after the laser pulse.* In ultrapure Ge impurity breakdown was observed 250 to 300 μsec after the laser pulse, and the breakdown threshold was independent of t_d.

Curve 2 of Fig. 13 shows the dependence of the impurity breakdown threshold on the duration of the microwave pulse. The threshold breakdown power becomes independent of τ for $\tau \geq 8$ μsec, and for $\tau < 1$ μsec impurity breakdown was not observed at all. The great difference between $P_d(\tau)$ for exciton breakdown as compared with impurity breakdown enables us to isolate one from the other by varying τ. For example, a 50-mW pulse of microwave power lasting less than 1 μsec produces only exciton breakdown (as is clear from Fig. 13).

Although the value of the impurity breakdown threshold varied from sample to sample, it was, as a rule, higher than the exciton breakdown threshold. Impurity breakdown was observed for temperatures of 1.3–4.2°K and the shape of the breakdown peak was essentially temperature independent.

Experiments using two successive pulses of equal durations and amplitudes revealed that impurity breakdown only takes place with the first pulse.

─────────

*We did not study the breakdown effect for longer delays.

Although we did not perform detailed studies of the impurity breakdown effect, one can conclude on the basis of the available experimental evidence that we are dealing with the so-called light-induced impurity breakdown which is observed in compensated samples [89]. The trailing edge of the breakdown peak describes the process of capturing electrons formed during breakdown by defects and other trapping centers. Based on the experimental results (see Fig. 16) we estimate the time to capture electrons at such centers in Ge to be about 10 μsec when T = 1.3°K.

We note that the impurity breakdown effect is much different from exciton breakdown in its characteristics. The great difference in the time required for breakdown to develop and the sharpness of the threshold and its dependence on the time delay between the microwave and laser pulses all make it very easy to distinguish between the two types of breakdown in an experiment.

3.4. Theory of Microwave Breakdown of Excitons.

Discussion of Experimental Results

The most striking characteristic of our experiments is the long delay between exciton breakdown and the exciting pulse. The optimal conditions (minimum threshold power) for exciton breakdown after the exciting laser pulse are produced in the sample after about 100 μsec following the production of the excitons (see Fig. 14). This is more than an order of magnitude longer than the lifetime of free excitons [90]. This fact suggests that microwave breakdown of excitons is in some way related to the presence of exciton complexes (either EHD or biexcitons) because their liftetimes are much longer than those of free excitons. In other words, the breakdown develops only in a system of electrons, excitons, and EHD, or in a system of electrons, excitons, and biexcitons. We shall discuss the experimental data in terms of these two models.

Let us begin with the electron-hole drop model since, as we shall see, it provides a natural explanation for the majority of the results. In this model we begin with the fact that most of the nonequilibrium carriers in the sample are found in drops with particle density n_0 and average radius $\bar{R}$. The concentration of drops is related to the carrier density induced in the sample by the following obvious equation:

$$\bar{n} = {}^4/_3 \, \pi \, \bar{R}^3 \, N n_0. \tag{2}$$

In addition to the drops the sample also contains excitons with density n and free carriers with density n_e. However, we have not taken these other configurations into account in Eq. (2) since $\bar{n} \gg n \gg n_e$ at the low temperatures of interest to us. During the first few microseconds following the excitation pulse dynamical equilibrium is established among these three groups of bound and free carriers. Free carriers combine into excitons and condense into drops, excitons dissociate into free carriers or are absorbed by drops, and drops capture and emit both excitons and free carriers. Each of the densities n, n_e, and $\bar{n}$ decreases with time because of recombination, but the equilibrium between them is preserved over an extended period.

No doubt the main cause of breakdown is the heating of free carriers by the microwave field. The amount of heating is easily estimated by the usual practice of equating the energy absorbed from the field to the losses experienced through phonon emission:

$$\frac{e^2 \mathscr{E}^2}{m\omega^2} \, \nu_{\text{ph}} = \frac{\hbar \bar{\omega} \nu_{\text{ph}}}{1 + 2N_{\bar{\omega}}} . \tag{3}$$

Here $\mathscr{E}$ is the field strength in the sample, ω is the frequency of the field, e and m are the charge and effective mass of the electron, ν_{ph} is the frequency of collisions with phonons, $\bar{\omega}$ is the average frequency of the phonons emitted and absorbed which, because of conservation of energy and momentum, is given by $\bar{\omega} = (1/\hbar)(2\bar{\varepsilon}ms^2)^{1/2}$; $\bar{\varepsilon}$ is the average electron energy, s is the speed of sound, and $N_{\bar{\omega}} = [\exp(\hbar\bar{\omega}/kT) - 1]^{-1}$ is the equilibrium phonon distribution function. Since $ms^2 \ll kT$ we find that for a small amount of heating $(\bar{\varepsilon} \sim kT)\ \hbar\bar{\omega} \ll kT$ and Eq. (3) gives

$$\bar{\varepsilon} = \frac{kT}{ms^2}\ \frac{e^2\mathscr{E}^2}{m\omega^2}. \tag{4}$$

On the other hand, for strong fields where $\hbar\bar{\omega} > kT$,

$$\bar{\varepsilon} = \frac{1}{2ms^2}\left(\frac{e^2\mathscr{E}^2}{m\omega^2}\right)^2. \tag{5}$$

We shall be most interested in the region where Eq. (5) is applicable since at breakdown $\bar{\varepsilon} \sim \mathscr{E}_0 \gg kT$ ($\mathscr{E}_0$ is the exciton binding energy). In this case $\bar{\varepsilon}$ increases very rapidly with the field $(\sim\mathscr{E}^4)$. We can easily estimate the fields necessary for everheating by substituting $\varepsilon \sim kT$ into Eq. (4). Then $e^2\mathscr{E}^2 \simeq (ms\omega)^2$ which, with $m \simeq 10^{-28}$ g, $s = 5 \cdot 10^5$ cm/sec, and $\omega = 6 \cdot 10^{10}$/sec, gives $\mathscr{E} \simeq 2$ V/cm. At fields of about 10 V/cm the average electron energy is about 10°K, which must correspond to the breakdown of the free excitons in the sample.

The key factor for explaining the experimental data discussed earlier is the reasonable assumption that the electron-hole drop, which is heated very little in the fields under consideration and is not destroyed by the free carriers bombarding it, still acts as an effective trapping center for electrons and holes. Let us estimate Γ_e^{-1}, the time for trapping at a single drop:

$$\Gamma_e^{-1} = [4\pi N\bar{R}^2 v_{\mathscr{E}}]^{-1} = \frac{n_0\bar{R}}{3\bar{n}v_{\mathscr{E}}}; \tag{6}$$

$v_{\mathscr{E}} = (\bar{\varepsilon}/m)^{1/2}$ is the mean random velocity of electrons in a microwave field of strength $\mathscr{E}$. By substituting the following numerical values into Eq. (6), $n_0 = 2 \cdot 10^{17}$ cm^{-3}, $\bar{n} \simeq 10^{15}$ cm^{-3} $\bar{R} \simeq 10^{-3}$ cm, and $v_{\mathscr{E}} \simeq 10^7$ cm/sec, we find that $\Gamma_e^{-1} \simeq 10^{-8}$ sec.

Thus, in a sufficiently pure sample or under conditions where other trapping centers (impurities or defects) are full after the exciting pulse, this mechanism is the determining factor in the concentration of free carriers. Electrical breakdown of excitons is determined by the ratio of the rate of free-carrier multiplication due to ionizing collisions $\alpha (\mathscr{E}) nn_e$ ($\alpha = \langle\sigma_e v_{\mathscr{E}}\rangle$ is the ionizing collision coefficient and σ_e is the cross section for destruction of an exciton by an electron) to the rate of capture by drops, and to the inverse process of combining into excitons βn_e^2. Therefore, the equations for changes in n and n_e are of the following form:

$$\begin{aligned} dn_e/dt &= \alpha\ nn_e - \beta\ n_e^2 - (\Gamma + \tau_e^{-1})\ n_e = \alpha\ n_e n - \beta\ n_e^2 - \tilde{\Gamma}_e n_e; \\ dn/dt &= -\alpha\ nn_e + \beta\ n_e^2 - \Gamma\ (n - n_{\scriptscriptstyle T}) - n/\tau_{ex}. \end{aligned} \tag{8}$$

The right sides of Eqs. (7) and (8) contain terms which take into account exciton recombination n/τ_{ex}, their capture by drops Γn, the evaporation of excitons from drops $\Gamma n_{\scriptscriptstyle T}$, and other mechanisms for electron capture n_e/τ_e unrelated to drops, so that overall $\tilde{\Gamma}_e = \Gamma_e + \tau_e^{-1}$. Clearly $\Gamma = (v_{\scriptscriptstyle T}/v_{\mathscr{E}})\Gamma_e$, where $v_{\scriptscriptstyle T}$ is the thermal speed of an electron. (We are assuming that both excitons and free carriers adhere to a drop when there is a collision with it.) The quantity $n_{\scriptscriptstyle T}$ characterizes the evaporation rate from a drop and is determined by the temperature of carriers in the drop. If the drop is not heated relative to the crystal lattice this quantity is equal to the exciton density in thermodynamic equilibrium with the drop at the temperature T. Equations (7) and (8) do not take into account thermal dissociation of excitons because of its low probability. Thermal dissociation does not play an important role in breakdown and is not really a source of free carriers because it is essentially a nonequilibrium process.

Since the coefficients α and β depend on the energy distribution of carriers, and therefore the field, α increases exponentially with $\mathscr{E}$, while β decreases. The coefficients $\tilde{\Gamma}_e$ and Γ are time dependent because carriers in the drop recombine with a time constant τ_0, and the total area of all drops decreases as $\exp(-\tfrac{2}{3}t/\tau_0)$. In the absence of an electric field the number of free carriers is small, and the solution to Eq. (8) for times just shortly after the exciting pulse $\sim (\Gamma + \tau_{ex}^{-1})^{-1}$ has a quasistationary form:

$$n(t) = \frac{\Gamma(t)\tau_{ex}}{1 + \Gamma(t)\tau_{ex}} n_{\mathrm{T}}. \tag{9}$$

When a pulse of field is applied at some point in time the criterion for breakdown is, of course, that $dn_e/dt > 0$, i.e., from Eq. (7)

$$\alpha n(t_0) - \beta n_e(t_0) - \Gamma_e(t_0) > 0. \tag{10}$$

Since breakdown develops over a very short time period, additional evaporation of excitons from the drop during this period can be neglected. Thus, to describe the breakdown peak the term $\Gamma(n - n)$ in Eq. (8) can be omitted and we may set $\tilde{\Gamma}_e(t) \simeq \Gamma_e(t_0) = \mathrm{const.}$ Then Eqs. (7) and (8) can be integrated to give

$$\frac{d\ln n_e}{dt} = \frac{\alpha\tilde{\Gamma}_e}{\alpha + \beta} \chi(n_e), \tag{11}$$

where the function $\chi(n_e)$ is determined by the transcendental equation

$$(n_e - n_{e\,\max})\frac{(\alpha + \beta)^2}{\alpha\tilde{\Gamma}_e} = -\chi + \ln(1 + \chi). \tag{12}$$

The constant of integration $n_{e\,\max}$ corresponds to the solution to Eq. (12) with $\chi = 0$ which, in view of Eq. (11), is equal to the maximum value of $\tilde{n}_e$ at the breakdown peak. It can be expressed in terms of the initial values $n(t_0) = \tilde{n}$ and $n_e(t_0) = n_e$ by comparing Eq. (7) with Eqs. (11)-(12) when $t = t_0$:

$$
\begin{aligned}
n_{e\,\max} &= \tilde{n}_e + \frac{d\tilde{\Gamma}_e}{(\alpha + \beta)^2}[\chi(n_e) - \ln(1 + \chi(n_e))] \\
&= \tilde{n}_e + \left[\frac{1}{(\alpha + \beta)\tilde{n}_e}\frac{dn_e}{dt} - \frac{\alpha\tilde{\Gamma}_e}{(\alpha + \beta)^2}\ln\left(1 + \frac{\alpha + \beta}{\alpha\tilde{\Gamma}_e}\frac{1}{n_e}\frac{dn_e}{dt}\right)\right]\bigg|_{t=t_0} \\
&= \tilde{n}_e + \frac{\alpha\tilde{n} - \beta\tilde{n}_e - \tilde{\Gamma}_e}{\alpha + \beta} - \frac{\alpha\tilde{\Gamma}_e}{(\alpha + \beta)^2}\ln\left[-\frac{\beta}{\alpha} + \left(1 + \frac{\beta}{\alpha}\right)\frac{\alpha\tilde{n} - \beta\tilde{n}_e}{\tilde{\Gamma}_e}\right].
\end{aligned}
\tag{13}
$$

The time development of the avalanche can be obtained by integrating Eq. (11):

$$t_{\max} - t_0 = \frac{\alpha + \beta}{\alpha\tilde{\Gamma}_e}\int_{\tilde{n}_e}^{n_{e\,\max}}\frac{dn_e}{n_e\chi(n_e)}. \tag{14}$$

The subsequent decay takes place with a time constant of $\tilde{\Gamma}_e/(1 + \beta/\alpha)$ since, as one easily sees from Eqs. (11)-(12) when $dn_e/dt < 0$ and $(n_{e\,\max} - n_e)[(\alpha + \beta)^2/\alpha\tilde{\Gamma}_e] \gg 1$, the function $\chi(n_e) \to -1$.

To simplify the following discussion and estimates we shall assume that $\beta n_{e\,max} \ll \Gamma_e$. This is usually the case since if processes quadratic in n_e are important in binding excitons, breakdown cannot be sharp. Therefore, assuming that $\beta/\alpha \ll 1$ Eq. (13) can be rewritten more compactly:

$$n_{e\,max} = \tilde{n}_e + \tilde{n}\left[1 - \frac{\tilde{\Gamma}_e}{\alpha\tilde{n}}\left(1 + \ln\frac{\alpha\tilde{n}}{\tilde{\Gamma}_e}\right)\right].\tag{15}$$

Near threshold, when $\alpha n_{e\,max} \ll \tilde{\Gamma}_e$, Eq. (12) reduces to

$$\chi^2 \simeq \frac{2\alpha}{\tilde{\Gamma}_e}(n_{e\,max} - n_e),\tag{16}$$

and Eq. (11) takes the form

$$\frac{d\ln n_e}{dt} = \sqrt{2\alpha\tilde{\Gamma}_e(n_{e\,max} - n_e)},\tag{17}$$

which can be solved to give

$$n_e = \frac{n_{e\,max}}{\cosh^2[\sqrt{2\alpha\tilde{\Gamma}_e n_{e\,max}}(t - t_{max})]}.\tag{18}$$

The breakdown peak in this case is nearly symmetric and its duration $(\alpha\tilde{\Gamma}_e n_{e\,max})^{-1/2}$ decreases with increasing $n_{e\,max}$, i.e., with the field. As breakdown is developing, when $\alpha\tilde{n} \gg \tilde{\Gamma}_e$, $n_{e\,max} \sim \tilde{n}$ and the expansion in Eq. (16) cannot be used. The peak becomes very asymmetric; the trailing edge decays, as already mentioned, with the time constant $\tilde{\Gamma}_e^{-1}$ but the leading edge increases much more rapidly in a time of the order of $\sim(\alpha\tilde{n})^{-1}$.

The solution just presented is valid over the time interval $\ll \Gamma^{-1} \propto (v_{\mathscr{E}}/v_T)\Gamma_e^{-1}$. For large times one cannot neglect in Eq. (8) the evaporation of excitons from the drop. This means that after the number of free carriers has greatly declined after breakdown, the number of excitons begins to slowly increase again and after a period of time of the order of $\sim\Gamma^{-1}$, which is one or two orders of magnitude longer than the duration of the breakdown peak, breakdown can be repeated. This agrees with the experimental results. The picture used here can account in a natural way for other breakdown regularities observed in the experiments. When $\beta \to 0$ one can use Eq. (9), the time dependence of both Γ and Γ_e, and the relationship between Γ and Γ_e to reduce the criterion for breakdown, Eq. (10), to this form:

$$\alpha n_T \tau_{ex} = \frac{v_{\mathscr{E}}}{v_T} + \frac{\tau_{ex}}{\tau_e} + 2\sqrt{\frac{v_{\mathscr{E}}}{v_T}\frac{\tau_{ex}}{\tau_e}}\cosh\left(\frac{2}{3}\frac{t - t_0}{\tau_0}\right),\tag{19}$$

where

$$t_0 = \frac{3}{2}\tau_0\ln\left(\sqrt{\frac{v_{\mathscr{E}}}{v_T}\tau_e\tau_{ex}}\,\Gamma_0\right),\tag{20}$$

$\Gamma_0 = \Gamma(t = 0)$ is the value of the parameter at the moment just following drop formation and is evidently proportional to $I^{2/3}$, where I is the intensity of the exciting pulse. Equations (19)–(20) correspond completely to the experimental results (see Figs. 12, 14, 15). By directly comparing Eq. (20) with Fig. 15 one finds the time constant τ_0 to be 35 μsec, which agrees well with EHD lifetimes obtained in other studies based on such phenomena as luminescence kinetics.

The presence of the first breakdown region for relatively short delay times $t_d < 40$ μsec is most easily explained by a weak heating of the sample following the excitation pulse. This heating increases exciton evaporation from the drop, thereby increasing $\tilde{n}$ in Eq. (10). We note that this heating can be very small ($< 1°K$) since n_T is exponentially dependent on the temperature.

Let us now discuss the dependence of the breakdown threshold on the duration of the microwave pulse. In all the preceding discussion we have assumed that the field does not change during the time the breakdown peak is developing. If, however, $\Gamma_e \tau < 1$, so that carriers are not trapped by the drop during the pulse, it is clear that the breakdown criterion becomes $\alpha \tilde{n} \tau \sim 1$, which means that each electron experiences at least one ionizing collision during the pulse duration period. Consequently, as τ decreases the ionization coefficient α corresponding to initiation of breakdown must increase together with the breakdown threshold field. Experimentally (see Fig. 13a) this increase begins when $\tau < 0.5 \cdot 10^{-6}$ sec, which means that $\Gamma_e \gtrsim 2 \cdot 10^6$ sec^{-1}, a value which is in qualitative agreement with estimates based on the duration of the trailing edge of the breakdown peak $\Gamma_e \simeq 5 \cdot 10^6$ sec^{-1}. By assuming the average value of $\Gamma_e = (3\bar{n}/n_0\bar{R})v_\mathscr{E} \simeq 3 \cdot 10^6$ sec^{-1}, with $v_\mathscr{E} = 10^7$ cm/sec, $n_0 \simeq 2 \cdot 10^{17}$ cm^{-3}, and $\bar{n} = 2 \cdot 10^{13}$ cm^{-3} (the last number was computed by assuming that each incident photon creates one electron-hole pair which then decays after 100 μsec with a time constant of $\tau_0 = 35$ μsec), we estimate the average drop radius to be $\bar{R} \simeq 10^{-3}$ cm and their density to be $N \simeq 5 \cdot 10^4$ cm^{-3}. Both of these estimates agree in order of magnitude with direct measurements of N and R in light-scattering experiments [36-39]. Thus, the model in which EHD's act as trapping centers for free current carriers can explain the experimental features of microwave breakdown of excitons in Ge when $T \leq 2°K$.

If one assumes that the excitons are bound into biexcitons rather than into EHD, we cannot at present suggest a satisfactory alternative description of all the facts. In a previous work [18] we suggested a model in which we observed microwave breakdown of excitons and explained the breakdown delay by assuming that after the exciting pulse the electron collision frequency ν in Eq. (1) is determined by the electron–biexciton collisions and $\nu \gg \omega$. As the biexciton density drops we assumed that the frequency decreased and the amount of energy absorbed by an electron increased such that the greatest heating in a given microwave field was found when $\nu \simeq \omega$. However, our later experiments convinced us that this explanation contradicted the experimental results. The model predicts that when $\nu \gg \omega$, $\Delta\varepsilon'' \gg \Delta\varepsilon'$, but experimentally one observes just the opposite, i.e., $\Delta\varepsilon' \gg \Delta\varepsilon''$ (see Section 3.1). Moreover, when $\nu \gg \omega$, $\Delta\varepsilon' \sim n_e/\nu^2 \sim n_e/n_B^2$ and should increase with time as n_B decreases. Direct cyclotron-resonance measurements of ν [41] also indicate that the opposite case is true when $\nu \ll \omega$.

Attempts to explain the breakdown delay by assuming, in analogy with the EHD model used above, that biexcitons trap free carriers also lead to difficulties even if such negative-ion complexes do in fact exist. The fact is that the ionization energy of a biexciton (the energy required to separate it into an electron, a hole, and an exciton) is at least a factor of 2 greater than that of an exciton, whereas the mean electron energy increases with the field as $\mathscr{E}^4$. Therefore the breakdown field for biexcitons cannot be more than 20% greater than that for excitons. But in fact biexcitons begin to break down sooner, because under the experimental conditions their density is orders of magnitude greater than the exciton density. The condition for biexciton breakdown is $\alpha_B n_B n_e - \gamma n_B n_e > 0$, where α_B is the biexciton collision ionization coefficient, n_B their density, and γ is a factor which determines the rate at which electrons are trapped by biexcitons. This relationship reduces to the obvious condition that $\alpha_B > \gamma$ and does not depend on the density n_B. It is therefore independent of time and, as such, cannot explain the delay of breakdown.

Another equally important shortfall of the biexciton gas model of breakdown is the fact that it provides no clear explanation for the fact that the breakdown peak develops over a very short period of time (< 1 μsec) and then does not continue even when the microwave pulse is on for long periods in single-shot experiments, or when the microwave pulse is repeated after ~10 μsec in two-shot experiments.

Thus, the electron-hole drop theory for microwave breakdown of excitons explains all the experimental facts cited (the presence and location of the minimum in the dependence of the breakdown threshold on the microwave pulse delay time, the shape of the breakdown peak, the dependence of the threshold power on the duration of the microwave pulse and on the power of the exciting light pulse, and the possibility of repeated breakdown only after a sufficiently long time), whereas the assumption of biexcitons cannot account for the observed effects in the breakdown.

CONCLUSION

We have presented results of studies of the kinetics of microwave conductivity in Ge using pulsed laser excitation and results for the microwave breakdown of excitons in both the continuous and pulsed regimes.

1. For temperatures $T \leq 3°K$ the microwave conductivity signal in Ge is the sum of two exponentials having time constants of about 1 μsec and 20-50 μsecs. The signal is due to free carriers in equilibrium with excitons and electron-hole drops (the fast exponential accounts for exciton trapping by EHD's and the slow exponential reflects the kinetics of recombination in the drop). It is observed that the slow exponential appears in the microwave conductivity signal when the excitation level corresponds to an initial electron-hole pair density of $\bar{n} \gtrsim 10^{13}$ cm^{-3}.

2. Breakdown of excitons in a microwave field has been observed. The threshold power for breakdown is about 5 mW and the effect is observed for temperatures of 1.3-2.5°K.

3. It has been demonstrated that a characteristic of the microwave breakdown of excitons in the continuous regime is a delay of breakdown relative to the exciting laser pulse.

4. Detailed studies of exciton breakdown by pulsed microwave fields reveal the following characteristics:

   a. Breakdown due to the microwave pulse is delayed by $t_d = 20$-180 μsec relative to the exciting laser pulse. Within this range of t_d there is a minimum in the breakdown threshold whose location in time depends on the excitation intensity;
   b. The breakdown is somewhat avalanche in nature, i.e., it develops and subsides in a short time (~0.5 μsec) regardless of the duration of the microwave pulse;
   c. The breakdown threshold is a minimum when the sample is situated at a maximum of the microwave electric field, and is completely absent in the microwave magnetic field;
   d. When the breakdown effect is studied using two successive microwave pulses, a second breakdown is observed when the interval between pulses is not shorter than some minimum time (10-15 μsec).

5. We have presented a theory for the breakdown based on collision ionization of excitons in the presence of EHD's which act as electron traps. The theory explains all the observed characteristics of the effect rather well.

6. By comparing this theory with the experiments we can estimate the EHD parameters: Their density, radius, and lifetime are $N = 5 \cdot 10^4$ cm^{-3}, $R \simeq 10^{-3}$ cm, and $\tau_0 = 35$ μsec at $T = 1.3°K$ respectively. These results agree well with data obtained in optical experiments.

Thus the microwave method is very informative when applied to studies of the collective properties of excitons at high densities, and helps explain many of the parameters in the free-exciton−EHD system.

We wish to thank A. A. Manenkov and V. A. Milyaev for their constant help during all phases of this work. I am very grateful to L. V. Keldysh for his creative participation in the work and for many productive discussions. I should also like to thank A. M. Prokhorov, V. S. Vavilov, V. S. Bagaev, and T. I. Galkin for many useful discussions, and V. A. Sanin, A. S. Seferov, S. P. Smolin for assistance in conducting the experiments.

LITERATURE CITED

1. S. A. Moskalenko, Fiz. Tverd. Tela, 4:276 (1962).
2. I. M. Blatt, K. W. Boër, and W. Brandt, Phys. Rev., 126:1961 (1962).
3. L. V. Keldysh and Yu. V. Kopaev, Fiz. Tverd. Tela, 6:2791 (1964).
4. M. A. Lampert, Phys. Rev. Lett., 1:450 (1958).
5. S. A. Moskalenko, Opt. Spektrosk., 5:147 (1958).
6. L. V. Keldysh, Proceedings of The Ninth International Conference on the Physics of Semiconductors, Moscow, 1968, Nauka, Leningrad (1968), p. 1303.
7. L. V. Keldysh, Usp. Fiz. Nauk, 100:514 (1970).
8. L. V. Keldysh, in: Excitons in Semiconductors [in Russian], Nauka, Moscow (1971), p. 5.
9. V. M. Asnin, A. A. Rogachev, and N. I. Sablina, Fiz. Tekh. Poluprovodn, 5:802 (1971).
10. V. M. Asnin, B. V. Zubov, T. M. Murina, A. M. Prokhorov, A. A. Rogachev, and N. I. Sablina, Zh. Éksp. Teor. Fiz., 62:737 (1972).
11. V. M. Asnin, Yu. N. Lomasov, and A. A. Rogachev, Pis'ma Zh. Éksp. Teor. Fiz., 18:242 (1973).
12. Ya. E. Pokrovskii and K. I. Svistunova, Pis'ma Zh. Éksp. Teor. Fiz., 9:435 (1969).
13. V. S. Bagaev, T. I. Galkina, O. V. Gogolin, and L. V. Keldysh, Pis'ma Zh. Éksp. Teor. Fiz., 10:309 (1969).
14. V. S. Vavilov, V. A. Zayats, and V. N. Murzin, Pis'ma Zh. Éksp. Teor. Fiz., 10:304 (1969).
15. J. C. Hensel and T. G. Phillips, Proceedings of the Eleventh International Conference on the Physics of Semiconductors, Warsaw, 1972, p. 671.
16. Ja. Pokrovsky, Phys. Status Solidi (A), 11:385 (1972).
17. C. Benoit à la Guillaume and M. Voos, Phys. Rev., B7:1723 (1973).
18. A. A. Manenkov, V. A. Mulyaev, G. N. Mikhailova, and S. P. Smolin, Pis'ma Zh. Éksp. Teor. Fiz., 16:454 (1972).
19. L. V. Keldysh, A. A. Manenkov, V. A. Milyaev, and G. N. Mikhailova, Zh. Éksp. Teor. Fiz., 66:2178 (1974).
20. L. V. Keldysh, A. A. Manenkov, V. A. Milyaev, and G. N. Mikhailova, Proceedings of the Twelfth International Conference on the Physics of Semiconductors, Stuttgart, 1974, p. 76.
21. L. V. Keldysh and A. N. Kozlov, Zh. Éksp. Teor. Fiz., 54:978 (1958).
22. E. A. Hulleraas and A. Ore, Phys. Rev., 71:493 (1947).
23. R. R. Sharma, Phys. Rev., 170:770 (1968).
24. O. Akimoto and E. Hanamura, Solid State Commun., 10:253 (1972).
25. W. F. Brinkman, P. W. Anderson, and B. Bell, Phys. Rev. B8:1570 (1973).
26. N. F. Mott, Philos. Mag., 6:287 (1961).
27. Ja. Pokrovsky, A. Kaminsky, and K. Svistunova, Proceedings of the Tenth International Conference on the Physics of Semiconductors, Boston, 1970, Cambridge University Press, Cambridge, Mass. (1970), p. 505.
28. Ya. E. Pokrovskii and K. I. Svistunova, Fiz. Tekh. Poluprov, 4:491 (1970).
29. W. Brinkman, T. Rice, P. Anderson, and S. Chui, Phys. Rev. Lett., 28:961 (1972).
30. M. Combescot and P. Nozieres, J. Phys. C. Solid State Phys., 5:2369 (1972).
31. R. M. Westervelt, T. K. Lo, J. L. Staehli, and C. D. Jeffries, Phys. Rev. Lett., 32:1051 (1974).
32. G. A. Thomas, T. M. Rice, and J. C. Hensel, Phys. Rev. Lett., 33:219 (1974).
33. V. M. Asnin and A. A. Rogachev, Pis'ma Zh. Éksp. Teor. Fiz., 7:464 (1968).

34. M. Gurny and M. Gliksman, Solid State Commun., 7:11 (1972).

35. V. M. Asnin and A. A. Rogachev, Pis'ma Zh. Éksp. Teor. Fiz., 14:494 (1971).

36. Ya. E. Pokrovskii and K. I. Svistunova, Pis'ma Zh. Éksp. Teor. Fiz., 13:297 (1971).

37. N. N. Sibel'din, V. S. Bagaev, N. A. Penin, and V. A. Tsvetkov, Fiz. Tverd. Tela, 15:177 (1973).

38. V. S. Bagaev, N. A. Penin, N. N. Sibel'din, and V. A. Tsvetkov, Fiz. Tverd. Tela, 15:3269 (1973).

39. J. M. Workock, D. H. Olson, and K. L. Shaklee, Bull. Am. Phys. Soc., 18:301 (1973).

40. P. S. Gladkov, B. G. Zhurkin, and N. A. Penin, Fiz. Tekh. Poluprovodn., 6:1919 (1972).

41. P. S. Gladkov, Candidate's Dissertation, Physics Institute, Academy of Sciences of the USSR (1972).

42. T. Sanada, T. Ohyama, and E. Otsuka, Solid State Commun., 12:1201 (1973).

43. J. C. Hensel, T. G. Phillips, and T. M. Rice, Phys. Rev. Lett., 30:227 (1973).

44. V. S. Vavilov, V. A. Zayats, and V. N. Murzin, in: Excitons in Semiconductors [in Russian], Nauka, Moscow (1971), p. 32.

45. V. M. Asnin, A. A. Rogachev, and N. I. Sablina, Pis'ma Zh. Éksp. Teor. Fiz., 11:162 (1970).

46. C. Benoit à la Guillaume, F. Salvan, M. Voos, I. M. Laurant, A. Bonnot, C. R. Acad. Sci. Paris, 272B:236 (1971).

47. J. Hvan and O. Christensen, Solid State Commun., 15:929 (1974).

48. A. S. Alekseev, V. S. Bagaev, and T. I. Galkina, Preprint FIAN No. 60 (1972).

49. C. Benoit à la Guillaume, M. Voos, and F. Salvan, Phys. Rev. Lett., 27:1214 (1971).

50. V. S. Bagaev, T. I. Galkina, N. A. Penin, V. B. Stopachinskii, and M. N. Churaeva, Pis'ma Zh. Éksp. Teor. Fiz., 16:120 (1972).

51. A. S. Alekseev, V. S. Bagaev, T. I. Galkina, O. V. Gogolin, N. A. Penin, A. N. Semenov, and V. B. Stopachinskii Tr. FIAN, 67:109 (1973).

52. V. N. Murzin, V. A. Zayats, and V. L. Konenko, Proceedings of the Eleventh International Conference on the Physics of Semiconductors, Warsaw, 1972, p. 678.

53. K. Betzler, B. Zhurkin, A. Karuzskii, and B. Balter, Preprint FIAN No. 71 (1975).

54. A. R. Hardtman and R. H. Rediker, Proceedings of the Tenth International Conference on the Physics of Semiconductors, Boston, 1970, p. 202.

55. R. S. Markiewicz, J. P. Wolf, and C. D. Jeffries, Phys. Rev. Lett., 32:1357 (1974).

56. J. P. Wolfe, W. L. Hansen, E. E. Haller, R. S. Markiewicz, C. Kittel, and C. D. Jeffries, Phys. Rev. Lett., 34:1292 (1975).

57. I. R. Haynes, Phys. Rev. Lett., 17:860 (1966).

58. V. S. Bagaev and L. I. Paduchikh, Fiz. Tverd. Tela, 13:484 (1971).

59. B. M. Ashkinadze, I. P. Krets, A. A. Pamrin, N. D. Yaroshetskii, Fiz. Tekh. Poluprovodn., 4:2206 (1970).

60. A. S. Kaminskii, Ya. E. Pokrovskii, and N. V. Alkeev, Zh. Éksp. Teor. Fiz. 59:1937 (1970).

61. L. V. Keldysh and A. P. Silin, Fiz. Tverd. Tela, 15:1532 (1973).

62. V. S. Bagaev, N. N. Sibel'din, and V. A. Tsvetkov, Pis'ma Zh. Éksp. Teor. Fiz., 21:180 (1976).

63. B. V. Zubov, V. P. Kalinushkin, T. M. Murina, A. M. Prokhorov, and A. A. Rogachev, Fiz. Tekh. Poluprov., 7:1614 (1973).

64. C. Benoit à la Guillaume, M. Capizzi, B. Etienne, and M. Voos, Solid State Commun., 15:1031 (1974).

65. J. C. Hensel, T. G. Phillips, and T. M. Rice, Phys. Rev. Lett., 30:227 (1973).

66. J. C. Hensel, Proceedings of the Ohi Seminar on Physics of Highly Excited States in Solids, Tomakomai, Japan, 1975.

67. P. S. Gladkov, B. G. Zhurkin, and N. A. Penin, Preprint FIAN No. 125 (1972).

68. A. A. Manenkov, V. A. Milyaev, G. N. Mikhailov, V. A. Sanina, and A. S. Seferov, Zh. Éksp. Teor. Fiz., 70:695 (1976).

69. Ya. E. Pokrovskii and K. I. Svistunova, Pis'ma Zh. Éksp. Teor. Fiz., 19:92 (1974).

70. Ya. E. Pokrovskii and K. I. Svistunova, Fiz. Tverd. Tela, 16:3399 (1974).

71. G. L. Bir and G. E. Pikus, Pis'ma Zh. Éksp. Teor. Fiz., 18:245 (1973).

72. A. S. Alekseev, T. I. Galkina, and N. A. Penin, Pis'ma Zh. Éksp. Teor. Fiz., 19:436 (1974).

73. Ya. E. Pokrovskii and K. I. Svistunova, Zh. Éksp. Teor. Fiz., 68:2323 (1975).

74. T. K. Lo, B. J. Feldman, and C. D. Jeffries, Phys. Rev. Lett., 31:224 (1973).

75. T. K. Lo, Solid State Commun., 15:1231 (1974).

76. G. A. Thomas, T. Phillips, T. Rice, and J. Hensel, Phys. Rev. Lett., 31:386 (1973).

77. T. K. Lo, B. J. Feldman, R. M. Westervelt, J. L. Staehli, C. D. Jeffries, and E. E. Haller, Preprint N UCB-34P20-167.

78. J. P. Wolfe, R. S. Markiewicz, and S. M. Kelso, Bull. Am. Phys. Soc., 20:471 (1975).

79. A. A. Manenkov, Doctoral Dissertation, Physics Institute, Academy Sciences of the USSR (1966).

80. N. V. Karlov and A. A. Manenkov, Quantum Amplifiers [in Russian], Izd. VINITI, Moscow (1966).

81. Yu. M. Kholinov, Report of Cryogenics Division, FIAN (1969).

82. L. D. Landau and E. M. Lifshits, Electrodynamics of Continuous Media [in Russian], Gostekhizdat, Moscow (1957).

83. R. N. Hall, and T. G. Solys, IEEE Trans. NS18:160 (1971).

84. B. V. Novikov, E. F. Gross, and M. A. Drygin, Pis'ma Zh. Éksp. Teor. Fiz., 8:15 (1968).

85. B. V. Zubov, A. A. Manenkov, V. A. Milyaev, G. N. Mikhailova, T. M. Murina, and A. S. Seferov, Fiz. Tverd. Tela, 18:706 (1976).

86. P. Suleebka and R. Snrau, J. Phys. D. Appl. Phys., 5:97 (1972).

87. Yu. K. Danileiko, A. A. Manenkov, A. M. Prokhorov, and V. A. Khaimov-Mal'kov, Zh. Éksp. Teor. Fiz., 58:31 (1970).

88. A. Macdonald, Microwave Breakdown in Gases, Wiley, New York (1966).

89. V. P. Dobrego, B. M. Konovalenko, and S. M. Ryvkin, Fiz. Tverd. Tela, 4:1911 (1962).

90. C. Benoit à la Guillaume, M. Voos, and F. Salvan, Phys. Rev., B5:3079 (1972).

MICROWAVE BREAKDOWN OF EXCITONS AND THE KINETICS OF FREE CARRIERS AND EXCITONS IN GERMANIUM IN THE PRESENCE OF ELECTRON-HOLE DROPS

A.A. Manenkov, V.A. Milyaev, G.N. Mikhailova, V.A. Sanina, and A.S. Seferov

The breakdown of excitons in Ge in a high-frequency electric field is studied. The highly sensitive microwave method for detecting free carriers following exciton breakdown enables us to study the kinetics of excitons at temperatures of 2.5-1.3°K. The presence of a constant exciton density for rather long periods of time is explained by a theory which takes into account the presence of a liquid excitonic phase in Ge, the electron-hole droplet. This model is supported by the fact that the temporal location of the minimum in the breakdown threshold does not depend on the frequency of the breakdown field. It is shown that the free carrier kinetics are due to a nonequilibrium mechanism, the ejection of electrons from the drop by Auger processes, and their capture by defects and impurities.

INTRODUCTION

The microwave breakdown of excitons has been observed and studied in detail [1, 2]. The following model is basic to explaining the experimental features of this type of breakdown. It is assumed that germanium held at liquid-helium temperatures contains a three-component system after excitation by a laser pulse; the three components are free carriers, free excitons, and electron-hole drops (EHD). The presence of EHD's has a significant impact on microwave breakdown because the drops act as trapping centers for free carriers. The theory developed in [2] enables one to use the experimental data to calculate the parameters of the EHD and the exciton gas, and these parameters are in good agreement with results obtained by other methods.

One of the principal characteristics of microwave breakdown of excitons in the presence of EHD's is the presence of a minimum in the threshold breakdown power after some delay between the laser pulse and the disruptive microwave pulse. This minimum is explained by noting that trapping of carriers by drops inhibits the collision ionization of excitons by free carriers in the microwave field during the intial stages. The minimum in the breakdown threshold corresponds to that delay time when the total drop surface area decreases (because of evaporation from the drop and recombination within it) until the drops become effective trapping centers.

It has been proposed [2] that equilibrium (gas−liquid) is maintained in the exciton-drop system while carrier recombination and drop evaporation take place. That is, at a fixed temperature the exciton density around the drop is, for all practical purposes, a constant as long as the drop exists.

The theory [2] predicts a very significant feature of microwave breakdown in the presence of an EHD; the temporal location of the minimum breakdown threshold is independent of the microwave frequency. We note that one can explain the observed minimum threshold in terms of an exciton breakdown model which does not include EHD (such a model for the breakdown of a pure exciton gas was discussed in [1]) by setting the field frequency equal to a time-dependent effective collision frequency for free carriers. But with such a model the location of the minimum breakdown threshold must be dependent on the field frequency [3].

As a consequence, it is of interest to study the frequency dependence of the breakdown threshold and to experimentally verify the constancy of the equilibrium exciton density during recombination of excited carriers in Ge.

In the present work we have investigated exciton breakdown in Ge using high-frequency electric fields (frequencies of the order of 300-900 MHz). A 10-MHz microwave spectrometer was used to detect the free carriers in the sample. The use of a two-frequency approach to measure breakdown threshold under a given set of experimental conditions (identical optical excitations, in particular) permitted a highly accurate study of the frequency dependence of exciton breakdown, especially the location of the minimum breakdown threshold.

In addition to microwave breakdown and exciton kinetics, we have studied free carrier kinetics in the presence of EHD's as a guide to the role of various processes in the decay of a drop, evaporation and Auger recombination in particular. Microwave methods of detecting free carriers have a number of advantages (high sensitivity, no contacts) which, when combined with pulsed breakdown of excitons, enable one to easily measure small exciton densities and to study their kinetics at densities ($\sim 10^{12}$ cm^{-3}) where optical methods simply lack sufficient sensitivity.

1. EXPERIMENTAL APPARATUS

Figure 1 shows a block diagram of the appratus used for pulsed microwave breakdown of excitons in Ge in the presence of high levels of optical excitation.

The germanium sample (Fig. 2) is situated in a flat-plate capacitor to which high-frequency (300-900 MHz) electric fields are applied in pulses to break down the excitons. The capacitor is placed either in a piece of short-circuited waveguide or in a reflecting rectangular cavity in the 10-MHz microwave spectrometer. In either case one records the changes in reflected microwave power due to changes in sample conductivity. The waveguide version of the microwave spectrometer is used to observe exciton breakdown caused by the high-frequency

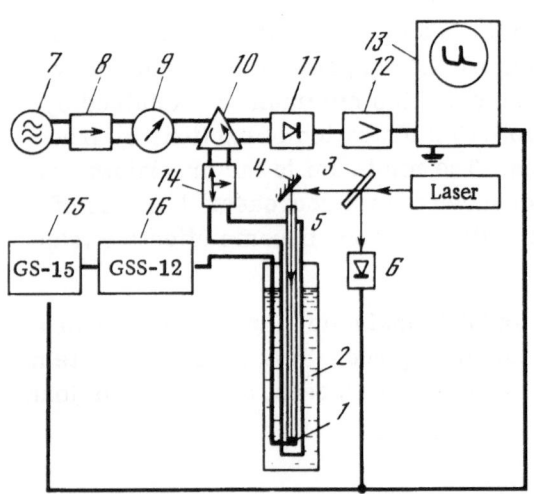

Fig. 1. Block diagram of apparatus. 1, sample holder; 2, cryostat; 3, beam splitter; 4, mirror; 5, light pipe; 6, photodiode; 7, klystron; 8, ferrite rectifier; 9, attenuator; 10, ferrite circulator; 11, microwave detector; 12, wide-band amplifier; 13, oscilloscope; 14, impedance transformer; 15, square-wave pulse generator; 16, microwave generator.

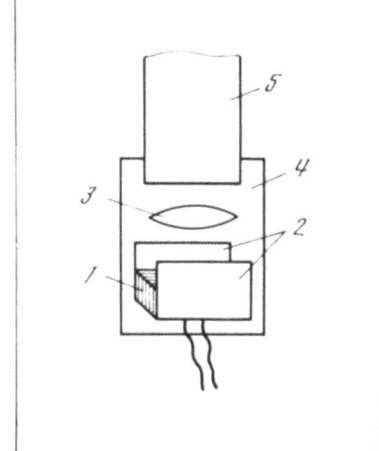

Fig. 2. Sample mount. 1, germanium sample;
2, capacitor plates; 3, lens; 4, Teflon holder;
5, light pipe.

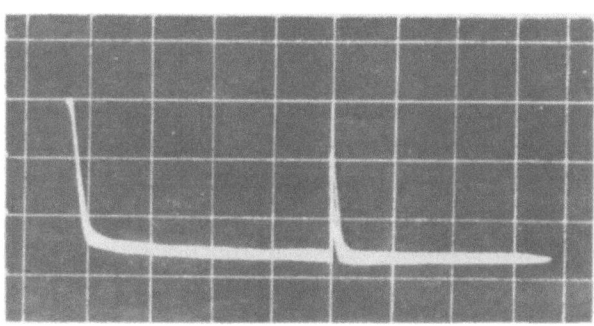

Fig. 3. Oscilloscope trace of the microwave
conductivity signal from a germanium sample.
Time zero corresponds to the laser pulse, and
after a delay the peak due to exciton breakdown appears. Abscissa scale is 20 μsec/div.,
ordinate is uncalibrated.

field. This preserves the amplitude linearity of the spectrometer relative to the recorded
reflection signal due to changes in sample conductivity which is proportional to carrier concentration. By selecting the phase of the reflected wave (by means of a short-circuiting plunger in the tuning circuit) one can record either the total complex conductivity of the sample, or
just its real part which depends on the absorption of microwave power.* The cavity version
of the microwave spectrometer is used in experiments on the simultaneous observation of
breakdown at two frequencies.

A Nd^{3+}YAG laser (λ =1.06 μm) was used for optical excitation of the germanium samples.
The Ge samples were 2 × 2 × 4 mm and had impurity concentrations of $N_A - N_B \sim 10^{10}$ cm^{-3}.

2. EXPERIMENTAL RESULTS AND DISCUSSION

The microwave conductivity signals due to the kinetics of free carriers generated by the
laser pulse take the form of the sum of two exponentials with time constants of 2 and 50 μsec.
The faster exponential accounts for binding of carriers into excitons and the excitons into
drops, and the slower one accounts for drop kinetics [2].

When an additional pulsed high-frequency field (300-900 MHz) of 1-10 μsec duration is
applied, exciton breakdown is observed, as indicated by a sharp spike in the conductivity (Fig. 3).
The breakdown has a threshold which depends on the strength of the high-frequency field. The
minimum breakdown threshold is observed when the high-frequency pulse is delayed relative
to the laser pulse by a time t_d which, as in microwave breakdown [2], depends on the excitation intensity. For excitation levels corresponding to carrier concentrations of the order of

---

* Reference 4 gives a detailed analysis of the method for measuring microwave conductivity
  in semiconductors.

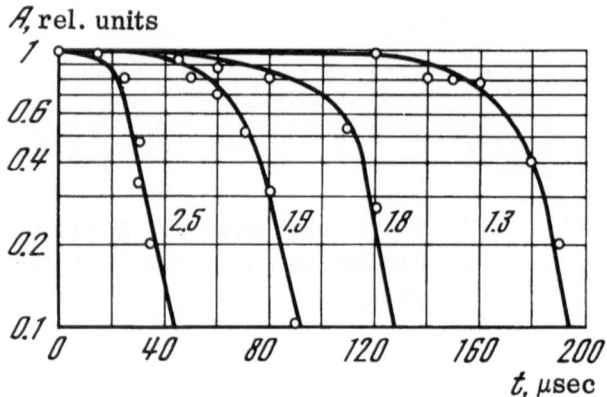

Fig. 4. Experimental dependence of the break-down peak amplitude on the delay of the break-down peak for various temperatures. The maximum amplitude is normalized to unity. The temperatures in degrees Kelvin are shown near each curve.

10^{15} cm^{-3} t_d is typically 100 μsec. It was observed that, other conditions being equal, the time location of the minimum breakdown threshold is independent of the frequency of the disruptive field. The generator frequency was varied from 300 to 900 MHz, and we also investigated microwave breakdown at 10 GHz.

We studied the amplitude of the breakdown peak as a function of the time delay for the disrupting pulse when breakdown is well developed, i.e., when the high-frequency field strength is much greater than threshold. It was found that the amplitude of the breakdown peak is constant over a wide range of delay times (Fig. 4). The range increases as the sample temperature drops. For large delays the breakdown peak amplitude is exponentially dependent on t_d with a time constant of 8 μsec.

Our experimental results are easily explained qualitatively by assuming the presence of a system composed of free carriers, an exciton gas, and electron-hole drops in the sample after laser excitation. In fact, the amplitude of the breakdown peak reflects the density of free carriers formed as a result of destruction of excitons. Thus, when breakdown is well developed and all the excitons have been destroyed, the amplitude of the breakdown peak gives information about the density of the exciton gas in the sample.* As in any liquid–vapor system, at a given temperature the density of excitons found in phase equilibrium with EHD's must be constant so long as the liquid phase exists. After the drops disappear because of recombination and evaporation, the electron density decreases with a time constant equal to the lifetime of free excitons.

Let us consider in greater detail the kinetics of the drops, excitons, and free carriers. They can be described by the following equations [5-7]:

$$\frac{dR}{dt} = -\frac{R}{3\tau_0} + \frac{v_{\mathrm{T}}}{n_0}(n - n_{\mathrm{T}}); \tag{1}$$

$$\frac{dn}{dt} = -\frac{n}{\tau_{ex}} - 4\pi R^2 N_d v_{\mathrm{T}}(n - n_{\mathrm{T}}) + \beta n_e^2; \tag{2}$$

$$\frac{dn_e}{dt} = -\frac{n_e}{\tau_e} - 4\pi R^2 N_d (v_{e\mathrm{T}} n_e - \Delta w_A n_0) - \beta n_e^2. \tag{3}$$

Here R and τ_0 are the EHD radius and lifetime; n, v_t, and τ_{ex} are the density, thermal velocity, and lifetime of the free excitons, respectively; n_{eT}, v_{eT}, and τ_e are the density, thermal velocity and lifetime of the free carriers (τ_e is related to impurity trapping, capture by defects, and so

_____

*We remark that the complete destruction of the exciton gas leads to the rather interesting physical situation in which the EHD is in an exciton vacuum.

on); N_d is the density of EHD; $n_0 \simeq 2 \cdot 10^{17}$ cm^{-3} is the carrier density in a drop; β is the cross section for binding carriers into excitons; w_A is the probability of Auger recombination of carriers in an EHD; Δ is the thickness of the EHD surface layer from which Auger electrons are derived; n_T is the thermal equilibrium density of excitons evaporated from a drop:

$$n_T = g \left(\frac{MkT}{2\pi\hbar^2} \right)^{3/2} e^{-\psi/kT}, \qquad (4)$$

where g is the multiplicity of the exciton ground-state degeneracy, M is the density-of-states effective mass for an exciton, and ψ is the work function for excitons leaving a drop.

Equation (1) gives the time dependence of the drop radius. The first term on the right is the rate of change of R due to carrier recombination inside the drop (both radiative recombination and Auger processes); the characteristic time is τ_0. The second term describes two processes, the thermal evaporation of excitons from the drop ($\propto n_T$) and the reverse current of excitons into the drop ($\propto n$). The second term can be neglected at low temperatures, and we obtain a solution for R(t) which is a simple exponential, just as in the case of exciton breakdown [2].

The first term of Eq. (2) for the density of free excitons accounts for exciton recombination, the second for evaporation and capture of excitons by drops (taking into account the density of drops within the sample). The third term in Eq. (2) describes binding of free carriers into excitons.

The first and second terms on the right side of Eq. (3) for free carriers account for electron capture by traps and crystal defects, and by EHD's. The third term describes emission of free carriers during Auger recombination. We assume that nonthermal emission of electrons in Auger processes takes place only from a rather thin (~0.1 μm) surface layer of the drop rather than from the entire drop volume as was assumed previously [8, 10]. Physically, the reason is very clear: Auger electrons produced well inside the drop lose their excess energy rapidly because of collisions and cannot reach the drop surface. Note that the slow exponential in the microwave conductivity signal has a 50-μsec time constant, which is longer than the characteristic decay time for the luminescence signal (~35 μsec) [2, 10]. This fact confirms that the microwave conductivity signal is due to Auger electrons emitted from the drop surface, while the luminescence signal arises from the recombination of electrons and holes inside the drop.

Thermal dissociation of excitons was not included in Eqs. (1)-(3) because it has such a low probability at low temperatures. We have also neglected thermal evaporation of carriers from the drop as compared with Auger processes. This assumption is valid because estimates of the equilibrium thermal density of free carriers from an equation similar to Eq. (4) (using ψ_e = 6 meV as the work function for emission from the drop) at T < 3.5°K give $n_{eT} \lesssim 10^4$ cm^{-3} whereas the density of Auger electrons is $n_{eA} \sim 10^{10}$-10^{11} cm^{-3}. The sensitivity of our microwave spectrometer allowed us to record free carrier densities $n_e \gtrsim 10^{10}$ cm^{-3}; as a consequence, the slow exponential in the microwave conductivity signal (Fig. 3) is due primarily to Auger electrons [8].

Analysis of Eq. (2) shows that the quasistationary solution for excitons (dn/dt = 0) corresponds to this value of n_{st}:

$$n_{st} = \frac{\beta n_e^2 + 4\pi R^2 N_d v_T n_T}{4\pi R^2 N_d v_T + \tau_{ex}^{-1}}. \qquad (5)$$

Clearly, when

$$\beta n_e^2 \ll 4\pi R^2 N_d v_T n_T,$$
$$\tau_{ex}^{-1} \ll 4\pi R^2 N_d v_T \qquad (6)$$

we have

$$n_{st} \approx n_T.$$

By assuming that $R_0 = 10^{-3}$ cm, $N_d = 5 \cdot 10^4$ cm^{-3}, $v_T = 3 \cdot 10^6$ cm/sec [2], $\tau_{ex} = 8 \cdot 10^{-6}$ sec, $n_T = 10^{12}$ cm^{-3} (T $\approx$ 1.5°K) [7], $w_A = 0.8 \cdot 10^4$ [9], and $\beta = 6 \cdot 10^{-4}$ cm^3/sec [10], we see that Eq. (6) is easily satisfied. The exciton density remains constant in time while R(t) does not become smaller than

$$R \propto \sqrt{\frac{\beta n_e^2}{4\pi N_d v_T n_T}} \simeq 1 \ \mu m. \tag{7}$$

This takes place 100-200 μsec after the exciting pulse. When the drop radius is small enough, Eq. (2) predicts that the exciton density exponentially decays with a time constant of τ_{ex}.

Let us now examine the free carrier kinetics [Eq. (3)]. Experimentally, the decay in the microwave conductivity signal indicates that the free carrier density changes substantially in about 10^{-4} sec whereas the characteristic times for carrier trapping by drops and impurities, for binding into excitons, and for Auger emission from drops are all of the order of 10^{-6}-10^{-7} sec. Therefore we should obtain a slowly varying solution for n_e by setting $dn_e/dt = 0$ in Eq. (3). When carrier trapping by impurities and defects is more important that binding into excitons [i.e., when βn_e^2 in Eq. (3) is negligible] the solution takes the form

$$n_e \approx \frac{4\pi R^2 \Delta N_d w_A n_0}{4\pi R^2 N_d v_{eT} + \tau_e^{-1}}. \tag{8}$$

Thus, it follows that when

$$\tau_e^{-1} > 4\pi R^2 N_d v_{eT}, \tag{9}$$

which is usually the case [2], the density of free carriers $n_e(t) \propto R^2(t)$, i.e., the density decays with a time constant $\propto \frac{3}{2}\tau_0$.

If combining of carriers to form excitons is the most important process then n_e must decay with a time constant of $3\tau_0$. The observed microwave signals show that in our samples the first case actually dominates, i.e., carrier capture by defects and impurities is the more probable.

CONCLUSION

Breakdown of excitons in a high-frequency electric field has been observed and studied. The use of high-frequency methods to detect free carriers due to exciton breakdown in a microwave spectrometer enables one to study exciton kinetics at temperatures as low as 1.3°K. The experimental data showing exciton densities to be constant over extended time periods are explained by a theory which takes into account the existence of a liquid exciton phase, the electron-hole drop, in germanium. This model is also supported by the observation that the temporal location of the minimum breakdown threshold is independent of the disrupting field's frequency. It is shown that the exponential decay of the free carrier density in the absence of exciton breakdown is due to Auger recombination of carriers inside the EHD.

We wish to thank A. M. Prokhorov and L. V. Keldysh for valuable discussions.

LITERATURE CITED

1. A. A. Manenkov, V. A. Milyaev, G. N. Mikhailova, and S. P. Smolin, Pis'ma Zh. Éksp. Teor. Fiz., 16:454 (1972).
2. L. V. Keldysh, A. A. Manenkov, V. A. Milyaev, and G. N. Mikhailova, Zh. Éksp. Teor. Fiz., 66:2178 (1974).
3. A. Macdonald, Microwave Breakdown in Gases, Wiley, New York (1966).
4. B. V. Zubov, A. A. Manenkov, V. A. Milyaev, G. N. Mikhailova, T. M. Murina, and A. S. Seferov, Fiz. Tverd. Tela, 18:706 (1976).
5. Ya. E. Pokrovskii and K. I. Svistunova, Fiz. Tekh. Poluprovodn., 4:491 (1970).
6. L. V. Keldysh, in: Excitons in Semiconductors [in Russian], Nauka, Moscow (1971), p. 5.
7. R. M. Westervelt, T. K. Lo, J. L. Shaehli, and C. D. Jeffries, Phys. Rev. Lett., 32:1051 (1974).
8. J. C. Hensel, T. G. Phillips, and T. Rice, Phys. Rev. Lett., 30:227 (1973).
9. K. Betzler, V. Zhurkin, A. Karuzskii, and B. Balter, Preprint FIAN No. 71 (1975).
10. P. S. Gladkov, Candidate's Dissertation, Physics Institute, Academy of Sciences of the USSR (1972).

MICROWAVE BREAKDOWN STUDIES OF EXCITON CONDENSATION IN Ge AND LUMINESCENCE DURING ONE- AND TWO-PHOTON EXCITATION OF CARRIERS

G.V. Zubov, A.A. Manenkov, V.A. Milyaev, G.N. Mikhailova, T.M. Murina, A.M. Prokhorov, and A.S. Seferov

We have investigated microwave breakdown of excitons and luminescence in pure Ge at 1.6°K when carriers are excited by lasers operating at 1.06 μm and 2.36 μm. It was determined that the characteristics of microwave breakdown do not depend on the excitation means for initial carrier concentrations of $5 \cdot 10^{13}$-$5 \cdot 10^{15}$ cm^{-3}. We observe that the fundamental electron-hole droplet parameters in the range of densities studied (lifetime, radius, and density) are identical for surface and bulk generation of carriers.

Recently there has been a great deal of interest in the condensation of excitons in Ge [1-10]. While most of the experimental work in this area has been with surface (single photon) excitation of carriers, the most important role in the condensation process is played by diffusion of carriers and excitons into the bulk of the sample. Thus, it is of interest to compare the condensation obtained with surface and bulk pumping because the conditions for forming the condensate can be different for the two cases. Bulk generation of carriers in Ge can be achieved by means of two-photon excitation with radiation from a Dy^{2+}CaF$_2$ laser [11].

We have used luminescence methods and microwave breakdown [12-15] of excitons to study the exciton condensation process in Ge for both single photon and two-photon excitation. It has already been demonstrated [12-15] that microwave breakdown of excitons is very dependent on the presence of electron-hole drops (EHD) in the exciton gas. Thus, we can use microwave breakdown of excitons to study both the condensation and the radius, density, and lifetime of EHD's. In order to compare the characteristics of the effect under one- and two-photon excitations, and to compare our results with those obtained using luminescence, we have studied microwave breakdown and recombination radiation from Ge using Nd^{3+}YAG ($\lambda_1 = 1.06$ μm) and Dy^{2+}CaF$_2$ ($\lambda_2 = 2.36$ μm) lasers for excitation purposes. Special precautions were taken to ensure that all experimental conditions were identical with the exception of the laser.

EXPERIMENTAL APPARATUS

The experimental apparatus diagrammed in Fig. 1 consists of a pulsed microwave spectrometer for studying the microwave photoconductivity of the sample and microwave breakdown of excitons, a pulsed infrared spectrometer for studying luminescence, pulsed Nd^{3+}YAG and Dy^{2+}CaF$_2$ lasers for carrier excitation in the Ge, and an optical helium cryostat.

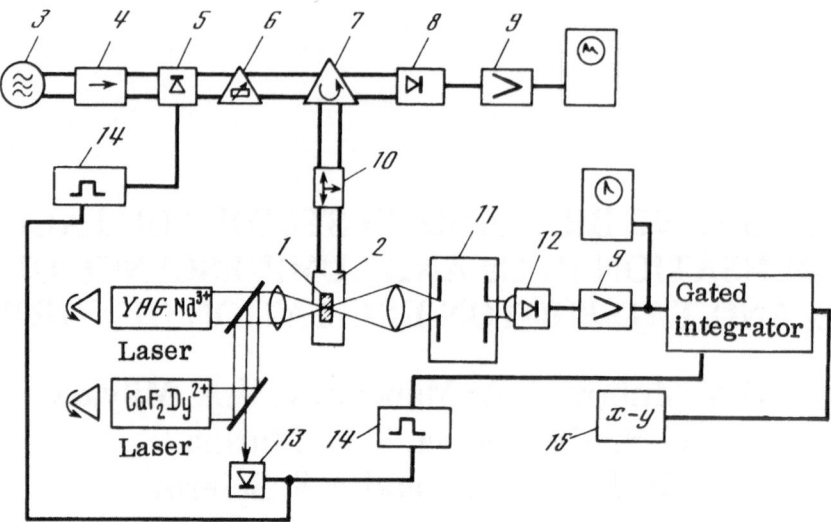

Fig. 1. Block diagram of apparatus for measuring the microwave breakdown of excitons and the luminescence from Ge due to laser excitation. 1, Ge sample; 2, microwave generator; 3, klystron; 4, ferrite isolator; 5, microwave modulator; 6, attenuator; 7, circulator; 8, microwave detector; 9, amplifier; 10, total impedance transformer; 11, MDR-2 monochromator; 12, FD-111 photodiode; 13, photodiode; 14, pulse generator; 15, recorder.

The 10-GHz microwave spectrometer is operated in the video amplifier mode with a pulsed output of 50 mW and a cw power of 0.1 mW, a time resolution of about 0.1 μsec, and a free carrier sensitivity of 10^{10} cm^{-3}. A detailed description of the instrument is found in [13].

The infrared spectrometer consists of an MDR-2 monochromator, an FD-111 photodiode, a gated integrator, a recorder, and an oscilloscope. The gate pulse is 1 μsec long with a resolution of 1 μsec.

The Nd^{3+}YAG and $Dy^{2+}CaF_2$ lasers produce giant pulses of 100 and 40 nsec with repetition rates of 50 and 500 Hz respectively.

We studied pure Ge samples containing 10^{10} cm^{-3} residual impurities. In order to reduce the surface recombination rates the Ge samples were etched in hot H_2O_2 + NaOH mixtures. The $5 \times 5 \times 2$ mm samples were fitted to a quartz light pipe inside a rectangular microwave cavity at the maximum electric field of the H_{102} oscillations excited in the cavity. The cavity was immersed in a liquid He optical cryostat. (Figure 2 shows a diagram of the lower part of the cryostat.) Measurements were performed at 1.6°K. A 3-mm-diameter, 15-mm-long quartz light pipe extracted the recombination radiation from the Ge.

Of great importance in exciton condensation studies is the exact value of the nonequilibrium carrier concentration because there is an on-going debate concerning the phase diagram and the threshold exciton densities, in particular, at which electron-hole drops are formed. Therefore we have made a special study to determine the carrier concentrations generated in one- and two-photon excitation. The most convenient and accurate method for measuring densities of carriers in our situation is to measure microwave absorption by the nonequilibrium current carriers (the method is described in detail in [16]).

Since the direct application of this method for determining the initial density of excited carriers is very difficult at helium temperatures because of exciton and EHD formation, the

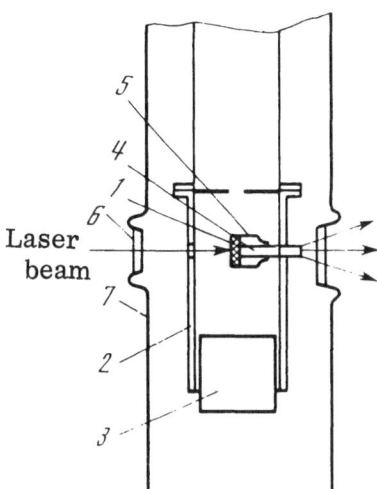

Fig. 2. Diagram of Ge sample inside microwave cavity. 1, Ge sample; 2, H_{102} microwave cavity; 3, plunger; 4, quartz light pipe; 5, Teflon sleeve; 6, quartz window; 7, helium bath cryostat.

density was determined at nitrogen temperature and the data extrapolated to helium temperatures taking into account the known temperature variations in the carrier diffusion coefficient [17]. In the two-photon version the density was determined from the radiation intensity in terms of the two-photon absorption coefficient [11].*

EXPERIMENTAL RESULTS AND DISCUSSION

The kinetics of microwave conductivity in Ge were studied using pulsed laser excitation at wavelengths of 1.06 μm and 2.36 μm with relatively low, constant levels of microwave power (p ~ 0.1 mW) which did not produce noticeable heating of carriers or breakdown of excitons. At T = 1.6°K the signal consists of two exponentials with time constants of τ_1 = 2 μsec and τ_2 = 50 μsec. As earlier reported [13] the microwave conductivity signal is due to free carriers in equilibrium with excitons and EHD's. The smaller time constant describes the process of carriers binding into excitons and exciton condensation, while the longer constant τ_2 = 50 μsec reflects the kinetics of decay in EHD's. It was observed that the shape of the microwave conductivity signals is identical for the two methods of excitation over a wide range of initial nonequilibrium current carrier densities, ranging from 10^{12} to $5 \cdot 10^{15}$ cm^{-3}. As the density increases, however, one observes a rapid distortion of the signal, evidently due to sample heating. This is also indicated by the fact that the same signal shape is observed when experiments are conducted at T = 4.2°K.

To break down the excitons a powerful microwave pulse is applied to the sample after a well-defined delay following the laser pulse. One observes a sharp spike in the conductivity which is due to collision ionization of excitons and their breakdown into free carriers. The fundamental characteristics of microwave breakdown were studied (in analogy with [13]), including the shape of the breakdown peak and its dependence on the time delay between the microwave and laser pulses, and the dependence on optical excitation level for both single- and two-photon carrier generation. These studies showed that the shape of the breakdown peak (length of the leading and trailing edges), the threshold power for breakdown, and the dependence of the threshold on the time delay of the microwave pulse are all identical for both single- and two-photon excitation and are the same as reported in [13].

Figure 3a shows the dependence of the temporal location of the minimum breakdown threshold on the light intensity. We see that the slopes of the curves for single-photon and two-

---

*We assume that the density found for two-photon excitation can be in error by a factor of 2 or 3.

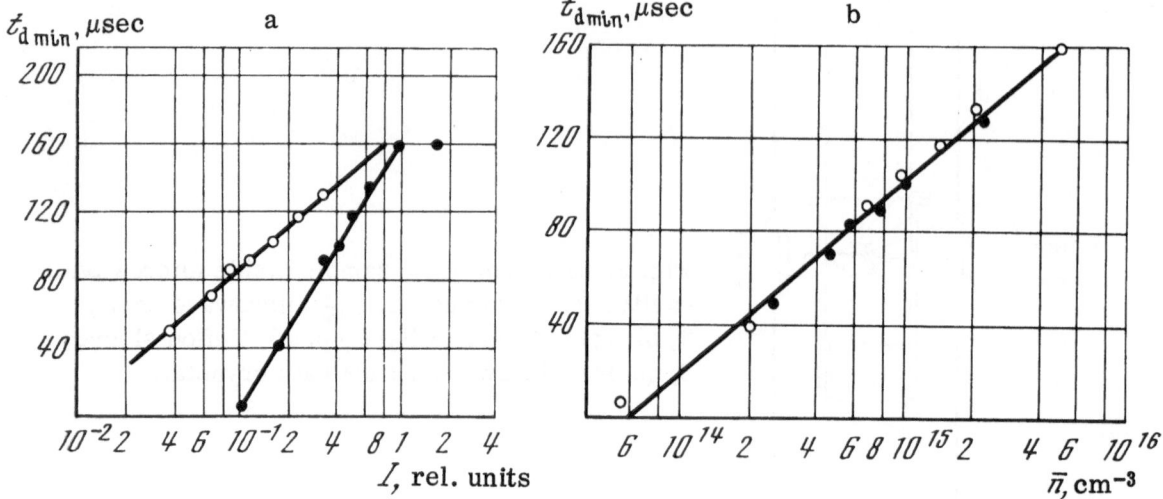

Fig. 3. Dependence of the minimum breakdown threshold delay $t_{d\,min}$ on the light intensity (a) and on the initial carrier density (b). Circles are one-photon excitation and dots are two-photon excitation.

photon excitation differ by a factor of 2. As we shall discuss in detail below, this fact is easily explained by the theory of microwave breakdown of excitons in the presence of EHD's [13] and is related to differences in carrier densities resulting from the light intensity used for the two modes of excitation. Figure 3b shows the delay $t_{d\,min}$ of the minimum threshold as a function of the initial density $\bar{n}$ of carriers induced in the sample. The time constant for this exponential dependence of $t_{d\,min}(\bar{n})$ is 35 μsec. One observes that the regions in which exciton breakdown exists in terms of both carrier density ($\bar{n} = 5 \cdot 10^{13}$ to $5 \cdot 10^{15}$ cm^{-3}) and in terms of time delay, for both surface and bulk excitation, coincide. Microwave breakdown was not observed beyond the carrier density limits shown here.

The kinetics of the emission line LA(709) were studied simultaneously with the microwave breakdown for both excitation modes. Figure 4 shows the time dependence of the line

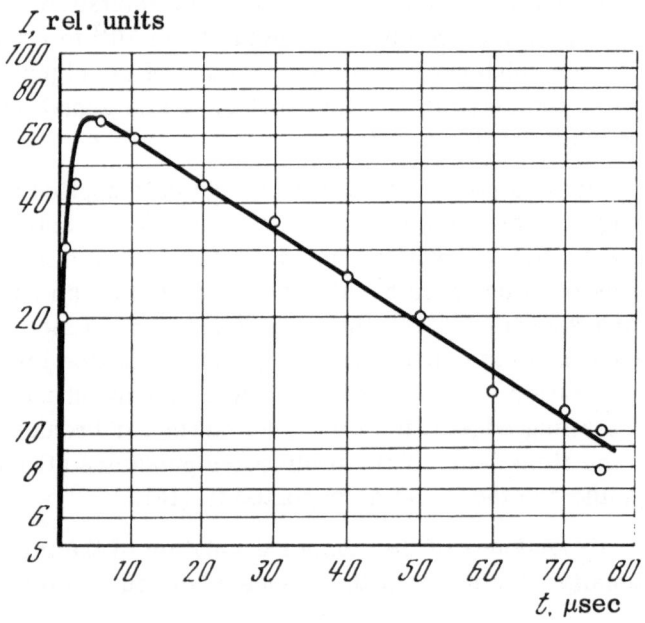

Fig. 4. Kinetics of the LA(709) line at 1.6°K for two-photon excitation, $\bar{n} = 10^{14}$ cm^{-3}.

intensity I(t) at T = 1.6°K for bulk carrier generation. The luminescence decays exponentially with a time constant of $\tau_0 = 35$ μsec. The luminescence decayed exponentially with the same τ_0 when the Nd^{3+}YAG laser was used for excitation. Measurements of the luminescence intensity as a function of light level used for excitation revealed that the threshold of the LA(709) line (defined as the sensitivity of the apparatus) is about 10^{12} cm^{-3} at 1.6°K, independent of the excitation mode used.

The most important experimental results are as follows:

1. Both the signal shapes for microwave conductivity and the recombination times are identical for one- and two-photon excitation.

2. Microwave breakdown of excitons in Ge is observed for both one- and two-photon excitation, and the characteristics are identical (time delay and signal shape). Exciton breakdown exists when the initial carrier concentration is in the range of $5 \cdot 10^{13}$ to $5 \cdot 10^{15}$ cm^{-3}. All the fundamental features of the effect are the same for the two modes of excitation when the densities are equal.

3. The LA(709) emission line appears when the excitation level corresponds to an initial carrier density of about 10^{12} cm^{-3}.

4. Both the luminescence time constant τ_0 and the characteristic time dependence of the minimum breakdown threshold delay on the initial carrier density are 35 μsec.

We shall discuss these results in terms of the exciton condensation model and the theory of microwave breakdown of excitons in the presence of EHD's. As already mentioned, the initial phase of the microwave conductivity signal describes the formation of EHD's which, experimentally, takes place with a characteristic time of about 2 μsec. That this time is independent of the mode of excitation (either one- or two-photon) indicates that the condensation process is the same for both surface and bulk generation of carriers.

It was shown in [15] that the slower exponential signal is due to free carriers surrounding the drop, carriers which are ejected from a thin surface layer of the EHD as a result of Auger recombination. The surface of an EHD is reduced exponentially with a time constant of $\frac{3}{2}\tau_0$, where τ_0 is the carrier lifetime inside an EHD; τ_0 is determined by the kinetics of luminescence. Thus, these experiments show that the lifetime of an EHD is independent of the means used to create carriers.

According to the microwave breakdown theory [13] the length of the trailing edge of the breakdown peak is determined by the capture of carriers produced in exciton breakdown by electron-hole drops. The characteristic trapping time is given by $\Gamma_e^{-1} \simeq n_0 \overline{R}/(3\overline{n}v_{\mathcal{E}})$, where Γ_e is the cross section for capture of carriers by a drop, $\overline{n}$ is the initial density of electron-hole pairs in the sample, n_0 is the carrier density in an EHD, and $v_{\mathcal{E}}$ is the average random electron velocity in the microwave field. Experimentally the trailing edge of the breakdown peak is about 0.2 μsec when $\Gamma_e \simeq 5 \cdot 10^6$ sec, and is the same for both single- and two-photon excitation. We then estimate that the radius of a drop with n = 10^{14} cm^{-3} (assuming that $v_{\mathcal{E}} = 3 \cdot 10^6$ cm/sec and $n_0 = 2 \cdot 10^{17}$ cm^{-3}) is $\overline{R} \simeq 10^{-3}$ cm. The drop density N_d is related to the average density of carriers induced in the sample by the obvious equation $N_d = 3\overline{n}/4\pi\overline{R}^3 n_0$. Substituting $\overline{R} \simeq 10^{-3}$ cm into this equation gives $N_d \simeq 10^5$ cm^{-3}. Since the breakdown peak is identical for the two excitation modes, the radius and density of an EHD are essentially the same for both surface and bulk carrier generation. This is evidently due to the effective diffusion of carriers and excitons during one-photon excitation [18].

The time delay of the minimum breakdown threshold is given by [13]

$$t_{d\,\min} = \tfrac{3}{2}\tau_0 \ln\left(\sqrt{\frac{v_{\mathcal{E}}}{v_{\mathrm{T}}}}\,\tau_e\tau_{ex}\Gamma_0\right),$$

where τ_0, τ_e, and τ_{ex} are the lifetimes of the drop, carriers, and excitations, v_δ is the electron velocity, v_T is the thermal velocity of the excitons, and $\Gamma_0 = 3\bar{n}v_\delta/n_0\bar{R}_0$ is the cross section for carrier trapping by drops at t = 0. Since $\bar{n} \propto I$ for one-photon excitation, and $\bar{n} \propto I^2$ for two-photon excitation, it is easy to understand the difference in the slopes of the function $t_{d\,min}$(I) in Fig. 3. We emphasize that the experimental value $\tau_0 = 35$ μsec for this dependence agrees well with the time constant for luminescence decay.

Let us now discuss the fact that microwave breakdown of excitons is observed for initial carrier concentrations of $\bar{n} = 5 \cdot 10^{13}$ to $5 \cdot 10^{15}$ cm^{-3} and is absent for densities outside this range. For densities greater than the upper limit ($\bar{n} > 5 \cdot 10^{15}$ cm^{-3}) breakdown is not observed because of sample heating. At low densities ($\bar{n} < 5 \cdot 10^{13}$ cm^{-3}) the mean density of the exciton gas in the sample in the presence of EHD's can be below the equilibrium value ($\sim 10^{12}$ cm^{-3} at T = 1.6°K [17]). Because of the low exciton density an electron will not experience even one ionizing collision during the period the microwave pulse is on. It is very difficult to break down the exciton gas under these conditions. This explanation is in qualitative agreement with the microwave breakdown theory for excitons, although more detailed experimental work is needed on the breakdown of excitons at low initial carrier densities.

Thus, there appears to be complete agreement between luminescence studies of EHD's and the microwave breakdown of excitons for both one- and two-photon excitation. Furthermore, the condensation process is the same for both types of excitation of carriers.

We wish to thank L. V. Keldysh for useful discussions and V. P. Kalinushkin and A. F. Shevel' for assistance in the experiments.

LITERATURE CITED

1. L. V. Keldysh, in: Excitons and Semiconductors [in Russian], Nauka, Moscow (1971), p. 5.
2. Ya. E. Pokrovskii and K. I. Svistunova, Pis'ma Zh. Éksp. Teor. Fiz., 9:435 (1969).
3. Ya. E. Pokrovskii and K. I. Svistunova, Fiz. Tekh. Poluprov., 4:491 (1970).
4. V. S. Bagaev, T. I. Galkina, O. V. Gogolin, and L. V. Keldysh, Pis'ma Zh. Éksp. Teor. Fiz., 10:309 (1969).
5. V. S. Vavilov, V. A. Zayats, and V. N. Murzin, Pis'ma Zh. Éksp. Teor. Fiz., 10:304 (1969).
6. N. N. Sibel'din, V. S. Bagaev, N. A. Penin, and V. A. Tsvetkov, Fiz. Tverd. Tela, 15:177 (1973).
7. J. C. Hensel, T. G. Phillips, and T. M. Rice, Phys. Rev. Lett., 30:227 (1973).
8. C. Benoit à la Guillaume, M. Voos, and F. Salvan, Phys. Rev., B5:3079 (1972).
9. T. K. Lo, B. J. Feldman, and C. D. Jeffries, Phys. Rev. Lett., 31:224 (1973).
10. J. P. Wolfe, R. S. Markiewicz, C. Kittel, and C. D. Jeffries, Phys. Rev. Lett., 34:275 (1975).
11. B. V. Zubov, T. M. Murina, B. R. Olovyagin, and A. M. Prokhorov, Fiz. Tekh. Poluprovodn., 5:636 (1971).
12. A. A. Manenkov, V. A. Milyaev, G. N. Mikhailova, and S. P. Smolin, Pis'ma Zh. Éksp. Teor. Fiz., 16:454 (1972).
13. L. V. Keldysh, A. A. Manenkov, V. A. Milyaev, and G. N. Mikhailova, Zh. Éksp. Teor. Fiz., 66:2178 (1974).
14. L. V. Keldysh, A. A. Manenkov, V. A. Miljaev, and G. N. Mikhailova, Proceedings of the Twelfth International Conference on the Physics of Semiconductors, Stuttgart, 1974, p. 76.
15. A. A. Manenkov, V. A. Milyaev, G. N. Mikhailova, V. A. Sanina, and A. S. Seferov, Zh. Éksp. Teor. Fiz., 70:695 (1976).
16. B. V. Zubov, A. A. Manenkov, V. A. Milyaev, G. N. Mikhailova, T. M. Murina, and A. S. Seferov, Fiz. Tverd. Tela, 18:706 (1976).
17. V. P. Aver'yanov, V. F. Bannaya, E. M. Gershenzon, and M. I. Ginzburg, Tr Gos. Nauchn.-Issled. Inst. Redk. Met., 25:169 (1968).
18. B. J. Feldman, Phys. Rev. Lett., 33:359 (1974).

MICROWAVE ABSORPTION BY NONEQUILIBRIUM CURRENT CARRIERS IN GERMANIUM. A METHOD FOR DETERMINING CARRIER CONCENTRATION

B.V. Zubov, A.A. Manenkov, V.A. Milyaev, G.N. Mikhailova, T.M. Murina, V.A. Sanina, and A.S. Seferov

We have measured the absorption of 3.2 cm microwave radiation by free carriers in germanium with pulsed laser excitation using two lasers: a Nd:YAG laser operating at 1.06 μm and a Dy:CaF$_2$ laser emitting at 2.36 μm. Two different types of absorption kinetics were observed at low and high excitation levels. When the density of nonequilibrium carriers $N < N_c$ the signal is exponential, but when $N > N_c$ the signal is clearly not exponential and exhibits an absorption maximum which is delayed by a time t_d relative to the exciting laser pulse. The observed signals are interpreted using a simple theory of microwave absorption in semiconductors. A method for measuring N_0 is presented which is based on measurements of the critical density N_c and the dependence of the delay time t_d on the intensity of the exciting radiation. Results of measurements of N_0 in germanium using this method are presented for both one- and two-photon excitation at 300°K and 77°K. The carrier diffusion length is estimated by comparing the shape of the microwave absorption signals for samples of different thicknesses.

One of the most important problems in the varied phenomena of semiconductor physics, and in exciton condensation in particular, is the task of precisely determining the number of nonequilibrium carriers generated by light. There are, at present, two principal methods for attacking this problem. The first is to measure the photoconductivity with a constant current [1] and the second involves free carrier absorption of infrared radiation [2, 3]. The first method is especially useful when carriers are generated in a steady-state manner. But when pulsed excitation is employed constant-current photoconductivity measurements are difficult because of photo-emfs and other contact phenomena.

Conductivity studies in the microwave region provide a more direct method for studying current carrier kinetics in the bulk of a semiconductor and enable one to determine the carrier concentration. The present paper reports work on the microwave conductivity of pure Ge crystals using pulsed laser excitation. The characteristic features of the kinetics reflected in the microwave absorption signal at different excitation levels, and their interpretation in terms of the theory of microwave absorption in semiconductors, permits us to proffer a new method for measuring nonequilibrium carrier densities which is distinguished by simplicity, reliability, and rather high accuracy. In addition, microwave conductivity studies provide a simple estimate of carrier diffusion lengths.

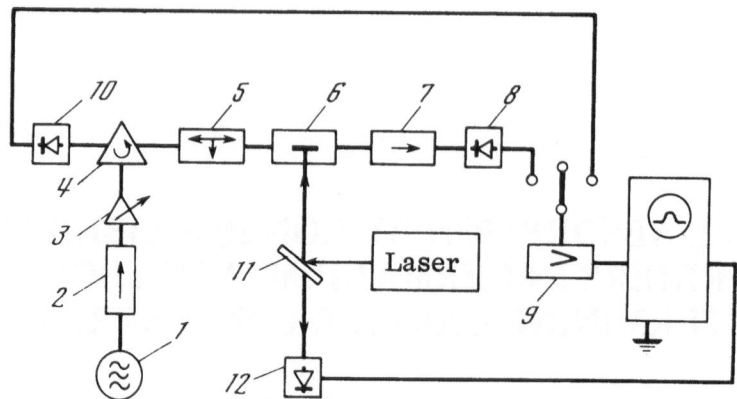

Fig. 1. Block diagram of the apparatus for studying microwave absorption at about 10 GHz in Ge. 1, klystron; 2, 7,
ferrite rectifiers; 3, attenuator; 4, ferrite circulator; 5,
impedance transformer; 6, Ge sample in waveguide; 8, 10,
microwave detectors; 9, amplifier; 11, mirror; 12, detector (photodiode).

EXPERIMENTAL APPARATUS AND MEASUREMENT METHOD

The conductivity of a semiconductor in a microwave field is a complex quantity:

$$\sigma = \sigma' + i\sigma''; \tag{1}$$

the real part σ' (the active conductivity) causes absorption of microwave energy, and the imaginary part (the reactive conductivity) produces phase-frequency shifts in the microwave circuit.
By using the well-known microwave methods (either waveguides or cavities) one can, in principle, determine both parts of σ and their variations during optical generation of carriers.
Cavity techniques offer greater sensitivity and efficiency when dealing with small samples and
low carrier densities. When studying larger samples having significant carrier densities it
is preferable to use waveguide methods since cavity techniques will induce strong nonlinearities in the signals.*

We have therefore selected the waveguide approach for our investigations of the microwave conductivity of Ge with optical excitation because we are interested in the conductivity
kinetics for a wide range of carrier densities (up to 10^{16} cm^{-3}) in relatively large samples
($2 \times 2 \times 0.2$-2 mm). Figure 1 shows a block diagram of the apparatus. The principal components are the 9.75-GHz klystron generator, a section of waveguide containing the Ge sample
which is excited by a laser pulse, a microwave receiver, and a recording oscilloscope. One
of two lasers was used for excitation purposes: either a Nd^{3+}YAG laser operating at 1.06 μm
or a Dy^{2+}CaF$_2$ laser operating at 2.36 μm. Both lasers operate in the giant pulse mode giving
pulses of 100 and 40 nsec duration with repetition rates of 50 and 500 Hz, respectively.

We studied plates of pure Ge (the impurity concentration did not exceed 10^{12} cm^{-3}) with
dimensions $2 \times 2 \times 0.2$, $2 \times 2 \times 1$, and $2 \times 2 \times 2$ mm at both room temperature and liquid-nitrogen
temperature. In order to reduce the surface recombination rate the samples were etched in
boiling H_2O_2 + NaOH.

---

* The analysis of cavity methods is completely analogous to the analysis used for EPR microwave spectroscopes [4].

The samples were placed in a rectangular waveguide so that the largest surface was parallel to the direction of the electric field. The laser beam entered through a slit in the narrow wall of the waveguide, and the laser beam was focused so that the entire sample was uniformly illuminated. The signal due to changes in the microwave conductivity of the sample during pulsed laser excitation is recorded from either changes in the power reflected from the sample (extracted by means of the ferrite circulator 4) or from the power transmitted to the detector 8. In the latter case the ferrite rectifier 7 isolates the detector from the sample and other circuit elements in order to preserve the independence of the circuit adjustment and the detector tuning. The total impedance transformer 5 is used for the initial adjustment of the equivalent resistance (or equivalent conductivity) induced in the waveguide line by the sample. As we shall now show, this adjustment allows us to record just the signal due to the active conductivity, which is of the greatest interest for analyzing the carrier kinetics and determining their density.

In the reflection scheme the signal is proportional to the change in reflected power ΔP_r, whereas in the transmission mode it is proportional to the sum $\Delta P_r + \Delta P_a$, where ΔP_a is the power change due to sample absorption. The power reflected from the sample is given by the well-known expression [5]

$$P_r = \Gamma\Gamma^*, \tag{2}$$

where $\Gamma = |\Gamma|e^{i\varphi}|\Gamma|$, φ are the modulus and phase of the reflection coefficient for the electric field:

$$|\Gamma| = \sqrt{\frac{(G-1)^2 + B^2}{(G+1)^2 + B^2}}, \tag{3}$$

$$\varphi = 2kd + \pi, \quad k = 2\pi/\lambda_B, \tag{4}$$

λ_B is the wavelength in the waveguide, d is the distance from the sample to the first minimum of the standing wave (it is assumed that a thin-plate sample whose thickness $2c \ll \lambda_B/2$ is situated in a transverse electric field in the waveguide), G and B are the active and reactive components of the equivalent parallel conductivity induced by the sample into the waveguide normalized to the characteristic waveguide conductivity $y_0 = 1/Z_0$, where Z_0 is the characteristic impedance of the waveguide,

$$G = \frac{r}{r^2 \sin^2 kd + \cos^2 kd}, \tag{5}$$

$$B = \frac{(r^2-1)\sin kd \cdot \cos kd}{r^2 \sin^2 kd + \cos^2 kd}, \tag{6}$$

r is the standing-wave coefficient.

It is clear from Eqs. (3)-(6) that in the general case where the sample has an arbitrary location along the guide, the reflected power is related to both the active and reactive components of the conductivity. Of course, when continuous excitation is employed it is quite simple to determine both components from standard measurements of the standing-wave coefficient SWC using measuring lines. Such measurements are more complex when dealing with conductivity kinetics and pulsed excitation. However, in our system it is very easy to exclude the reactive component by first adjusting the waveguide with the total-impedance transformer. By using this method practically any phase of the reflection coefficient can be established, including the values $\varphi = m\pi$ (m = 1, 2, 3) at which B = 0. Thus, one can easily create conditions for both the reflected and transmitted signals in which one is recording only the active conductivity, the part due to microwave power absorption.

The absorbed power is determined from the expression

$$P_a = \sigma' E_i^2, \tag{7}$$

where E_i is the field inside the sample and is dependent on the dielectric constant and sample shape. For thin rectangular plates with dimensions $2a \times 2b \times 2c$ ($a = b \gg c$) the internal field can be approximated by the equation (if the skin effect is absent) [6]

$$E_i = \frac{E_0}{1 + a(\varepsilon - 1)}, \tag{8}$$

where E_0 is the external field at the sample site, $\alpha = \pi c/4a$ is the depolarizing factor, and ε is the sample dielectric constant which is composed of the lattice contribution ε_0 and the dielectric susceptibility which is related to the current carriers:

$$\varepsilon = \varepsilon_0 + \varepsilon' + i\varepsilon'',$$
$$\varepsilon' = -\frac{4\pi Ne^2}{m(\omega^2 + \nu^2)} = -\frac{4\pi\sigma'}{\nu}; \tag{9}$$
$$\varepsilon'' = \frac{4\pi Ne^2}{m\omega}\frac{\nu}{\omega^2 + \nu^2} = \frac{4\pi\sigma'}{\omega}.$$

Here N is the free carrier concentration, ω is the frequency of the electromagnetic field, ν is the effective collision frequency, and e and m are the charge and effective mass of the carriers.

Note that Eq. (8) is valid for quasi-steady-state and uniform fields, i.e., for $\lambda_B/|\sqrt{\varepsilon}|$ larger than the sample dimensions. This condition is satisfied in our case. Then, in view of Eqs. (8) and (9), Eq. (7) becomes

$$P_a(\sigma') = \frac{\sigma' E_0^2}{(1 + a\varepsilon_0 - a - 4\pi a\sigma'/\nu)^2 + (4\pi\sigma' a/\omega)^2}, \tag{10}$$

where

$$\sigma' = \frac{Ne^2\nu}{m(\omega^2 + \nu^2)}.$$

EXPERIMENTAL RESULTS AND DISCUSSION

The primary goal of our work has been to study the kinetics of microwave conductivity in Ge using two methods of optically generating carriers. The first is the one-photon method at 1.06 μm in which the photon energy exceeds the band gap E_g; the second method is the two-photon approach at 2.36 μm wherein $\hbar\omega_2 < E_g$. A further aim is to determine the carrier concentration for these two methods.

Figure 2 shows oscilloscope photographs of microwave absorption signals in Ge for various levels of pulsed laser excitation at 300°K. While the 300°K results show considerable variations between thick and thin samples (Fig. 2e, f) the microwave conductivity signals at 77°K are identical for both excitation methods (using Nd^{3+} YAG and $Dy^{2+}CaF_2$ lasers) for all three samples (the dimensions are given above). This result is easily explained in terms of carrier diffusion which is important in the single-photon case where carriers are generated only in a thin surface layer. In the two-photon case the excitation is nearly uniform throughout the entire sample volume. Clearly, if the sample thickness is greater than the characteristic carrier diffusion length, the carriers will be nonuniformly distributed and one may observe distortions in the signal describing the carrier kinetics. We shall not dwell further on these distortions except to note that the results are in qualitative agreement with the known diffusion coefficients for free carriers in Ge at both 300°K and 77°K (60 and 1600 cm^2/sec, respectively).

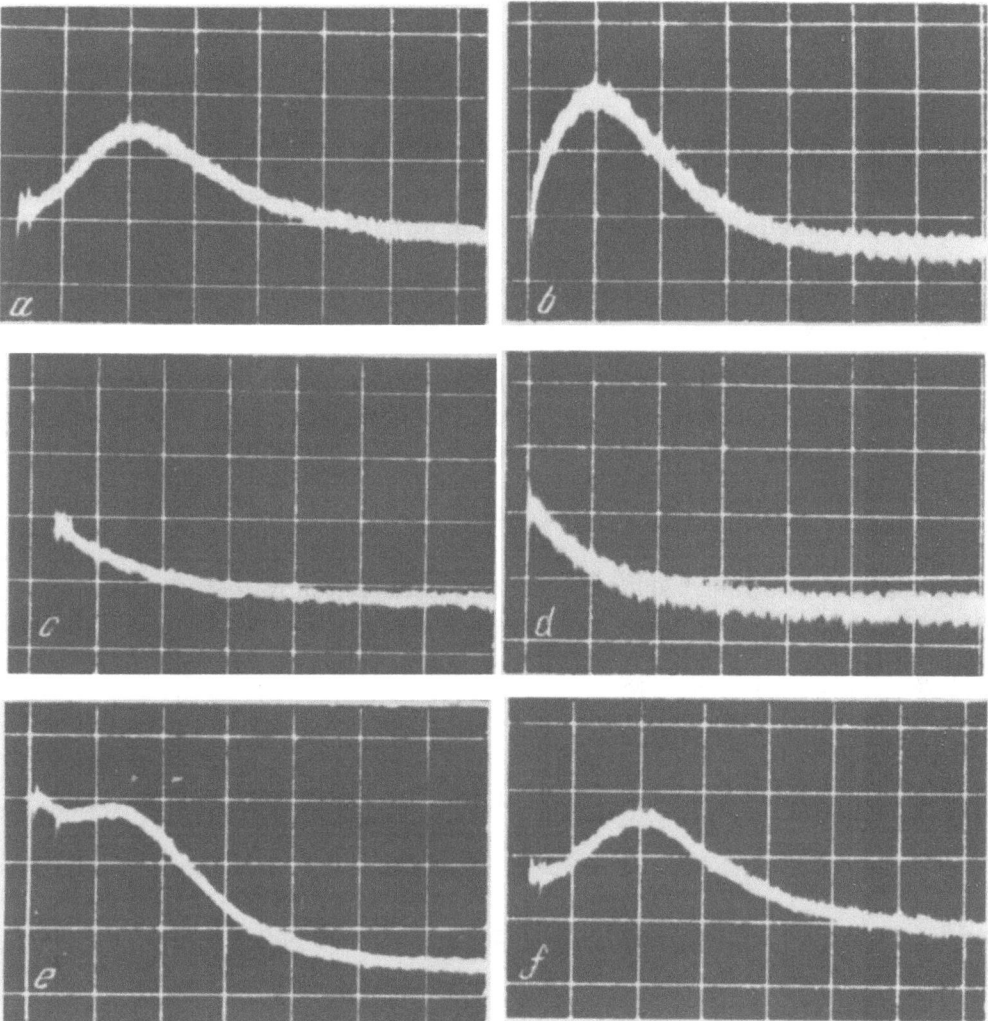

Fig. 2. Oscilloscope traces of microwave absorption in Ge for various levels of pulsed laser excitation at 300°K. On the left are one-photon excitation results; on the right are two-photon results. Traces a-d are from a sample with dimensions $2 \times 2 \times 0.2$ mm; traces e and f are from a sample with dimensions $2 \times 2 \times 2$ mm. Traces a, b, e, f were obtained with high levels of excitation $N_0 > N_c$, traces c, d with $N_0 < N_c$. Time scale is 100 μsec/div.

The most noticeable feature of the signals in both the single-photon and two-photon cases is the great difference observed for different intensities I of the exciting radiation. When $I < I_c$ (where I_c is a critical intensity which depends on the type of laser, either Nd^{3+}YAG or $Dy^{2+}CaF_2$, and the sample temperature) the signals are the usual exponentials, whereas when $I > I_c$ one observes signals in which the location of the maximum (i.e., the delay time t_d relative to the laser pulse) is a function of the excitation intensity rather than the maximum amplitude being such a function of the intensity. This result is easily understood in terms of the theory of microwave absorption in semiconductors. It follows from Eq. (10) that as a function of the free carrier density N the absorbed power has a maximum* (obtained from the condition

---

* This source for an explanation of the signal shape was suggested by L. V. Keldysh.

$\partial P_a /\partial N = 0$) when

$$N_c = \frac{m\omega^2}{4\pi e^2}\left(\varepsilon_0 + 1 + \frac{a}{c}\,\frac{4}{\pi}\right)\left[1 + \left(\frac{\nu}{\omega}\right)^2\right]^{1/2}. \tag{11}$$

Since we are dealing with nonequilibrium carriers in pulsed excitation, $N = N(t)$, both the critical density N_c and the corresponding maximum in microwave absorption are reached when $t = t_d$. The decrease in microwave absorption when $N > N_c$ indicates, from a physical point of view, that at such carrier densities the microwave field does not penetrate into the sample because of an increase in the effective dielectric constant or a metallization of the semiconductor.

The value of the time delay t_d of the maximum absorption is determined by the recombination of carriers. It follows from the results cited above that, at low carrier densities ($N < N_c$), the specific equation governing recombination is exponential in form:

$$N = N_0 e^{-t/\tau_0}, \tag{12}$$

where τ_0 is the carrier lifetime. Assuming that recombination follows the same law at high densities ($N > N_c$), one can establish a direct relationship between t_d, N_c, and the initial concentration N_0:

$$t_d = \tau_0 \ln (N_0/N_c). \tag{13}$$

For one-photon and two-photon generation of carriers we have, respectively,

$$N_0^1 = \beta_1 I_1, \quad N_0^2 = \beta_2 I_2^2,$$

where I_1 and I_2 are the intensities of the exciting laser pulses, and β_1 and β_2 are constants. Thus, for one-photon excitation

$$t_d = \tau_0 \ln (I_1/I_{1c}) \tag{14}$$

and for two-photon excitation

$$t_d = 2\tau_0 \ln (I_2/I_{2c}). \tag{15}$$

Here I_{1c} and I_{2c} are the critical laser pulse intensities at which $t_d = 0$ and $N_0 = N_c$.

Equations (12)-(15) provide a simple way for determining the initial density of nonequilibrium carriers if N_c is known; one simply measures the delay time of the maximum microwave absorption peak as a function of the intensity of the exciting laser pulse. The critical density can be computed from Eq. (11) if one knows either the collision frequency ν or the carrier mobilities $\mu_e + \mu_h$. These quantities are related as

$$\mu_e + \mu_h = \frac{e\nu}{m\,(\nu^2 + \omega^2)}.$$

On the other hand, since the ratio N_0/N_c can be determined directly from our microwave measurements, the method provides a new tool for determining the collision frequency ν if the density N_0 is known along with its dependence on some parameter like temperature.

The interpretation of the microwave absorption using pulsed excitation as presented here was carefully verified experimentally with germanium samples at 300 and 77°K. Figure 3a shows the dependence of the time delay of the peak in the microwave absorption signal on the

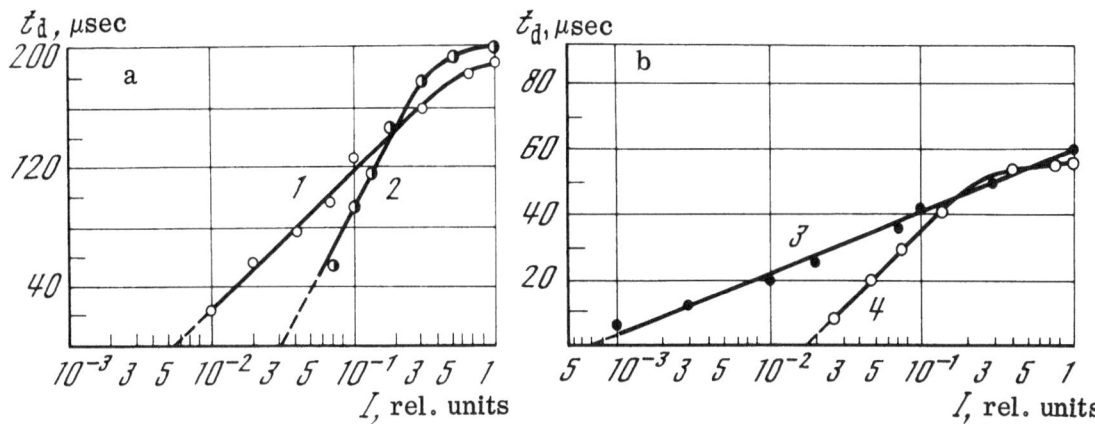

Fig. 3. Delay time of the microwave absorption peak as a function of the intensity of the laser excitation pulse. a) $T = 300°K$; b) $T = 77°K$. 1, 3) one-photon excitation: $\tau_0 = 42$ μsec, $\tau_0 = 8.5$ μsec; 2, 4) two-photon excitation: $2\tau_0 = 84$ μsec; $2\tau_0 = 17$ μsec. $2 \times 2 \times 0.2$ mm Ge sample.

intensity of both the Nd^{3+}YAG and $Dy^{2+}CaF_2$ lasers for a 0.2-mm-thick sample of Ge. It is clear that the results are well described by Eqs. (14) and (15) over the entire intensity range studied except for a narrow region near the maximum intensity $(0.3-1)I_{max}$. The deviations may be due to either the skin effect on the microwave field in the sample or to deviations from exponential behavior in the recombination law at high densities. The values of the nonequilibrium carrier lifetime τ_0 as determined from the slopes of the lines for $t_d(I)$ for single- and two-photon excitation are shown in Fig. 3, and there is good agreement in the values obtained.

The critical densities computed from Eq. (11) for a $2 \times 2 \times 0.2$ mm Ge plate with the known carrier mobilities are $N_c = 1.6 \cdot 10^{14}$ cm^{-3} at 300°K and $N_c = 1.2 \cdot 10^{13}$ cm^{-3} at 77°K.

With these values of N_c and the measured values of t_d and τ_0, one can use Eq. (12) to find the initial density N_0 at any excitation intensity where these equations are valid. It is clear from Fig. 3 that this procedure is correct up to the point where $N_0 \simeq 50N_c$ (i.e., where $t_d = 170$ μsec) at 300°K for both one-photon and two-photon excitation. The 77°K measurements resulted in similar dependences and features in the signals, but the time constant τ_0 is much smaller as one would expect at low temperatures (see Fig. 3b). The important result of the 77°K measurements is the observation that a 2-mm-thick Ge sample excited at the surface by 1.06 μm radiation is completely filled with nonequilibrium carriers after about 8 μsec.

The lifetime and density of free carriers can also be determined at other microwave frequencies (300-1000 MHz). The experimental arrangement is very simple (Fig. 4). It con-

Fig. 4. Diagram of apparatus for recording microwave absorption signals in Ge at frequencies near 1 GHz. 1, GSS-12 microwave generator; 2, impedance transformer; 3, measuring capacitor with Ge sample; 4, detector; 5, U3-7A wide-band amplifier; 6, oscilloscope; 7, photodiode; 8, laser.

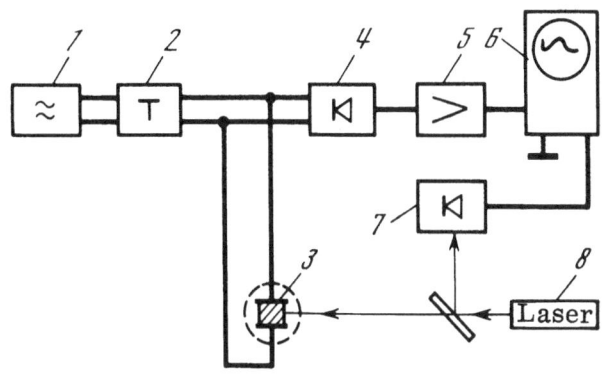

sists of a GSS-12 microwave generator, an impedance transformer, the measuring capacitor containing the sample, a U3-7A wide-band amplifier with a set of detector heads, and an oscilloscope.

Note that the critical density N_c depends on the field frequency ω according to Eq. (11). Therefore, it is of interest to investigate the conductivity of Ge at two frequencies simultaneously; we used frequencies near 1 GHz and 10 GHz. We used an arrangement similar to that described in [7] consisting of a microwave videospectroscope combined with a microwave circuit. It was observed that the absorption signals at the two frequencies had identical shapes for a given excitation level (the shapes are similar to those shown in Fig. 2). The only difference observed was a larger delay time t_d for the 1 GHz signal. This result is easily understood in view of Eq. (13) and Eq. (11):

$$t_d^1 - t_d^2 = \tau_0 \ln (N_C^2/N_C^1) = \tau_0 \ln (\omega^2/\omega^1), \tag{16}$$

where the superscripts 1 and 2 denote the frequencies 795 MHz and 10 GHz respectively.

In the experiment with $\tau_0 = 38$ μsec the following results were obtained: $t_d^1 = 250$ μsec at 795 MHz, $t_d^2 = 190$ μsec at 10 GHz. These results agree with Eq. (16) to within the experimental errors.

The main source of error in using this method for determining N_c and τ_0 is the uncertainty in measuring the delay time of the signal maximum. In our experiments this uncertainty is ± 15 μsec at 300°K and ± 5 μsec at 77°K. This introduces about 25% error into the density determination.

In conclusion, our measurements indicate that microwave absorption of nonequilibrium current carriers in semiconductors provides a direct, reliable method for obtaining the carrier density, the diffusion coefficient, and the effective collision frequency. We wish to thank A. M. Prokhorov and L. V. Keldysh for discussions.

LITERATURE CITED

1.   S. M. Ryvkin, Photoelectric Effects in Semiconductors [in Russian], Fizmatgiz, Moscow (1963).
2.   R. Bube, Photoconductivity of Solids, Wiley, New York (1960).
3.   V. S. Vavilov, Radiation Effects in Semiconductors [in Russian], Fizmatgiz, Moscow (1963).
4.   N. V. Karlov and A. A. Manenkov, Quantum Amplifiers [in Russian], VINITI, Moscow (1966).
5.   Theory of Microwave Transmission Lines [in Russian], Sov. Radio, Moscow (1951), Vol. 1.
6.   L. D. Landau and E. M. Lifshitz, Electrodynamics of Continuous Media [in Russian], Gostekhizdat, Moscow (1957), pp. 43, 65.
7.   A. A. Manenkov, V. A. Milyaev, G. N. Mikhailova, V. A. Sanina, and A. S. Seferov, Zh. Éksp. Teor. Fiz., 70:695 (1976).

MICROWAVE METHODS FOR STUDYING EXCITON CONDENSATION IN SEMICONDUCTORS

A.A. Manenkov

It is shown that study of the kinetics of microwave conduction following pulsed optical excitation of carriers and the measurement of microwave breakdown of excitons provide new exciting possibilities for investigating excitons in semiconductors. These possibilities are discussed in an analysis of a system of free carriers (FC), free excitons (FE), and electron-hole drops (EHD's) in a strong microwave electric field which causes collision ionization of the excitons. Experimental data are cited from studies of microwave conductivity and microwave breakdown of excitons in germanium and are used to determine some parameters of the FC−FE−EHD system (lifetime, equilibrium densities, EHD dimensions, free carrier and exciton cross sections for capture by drops, Auger process parameters, and so on).

There are now many methods being used to study exciton condensation in semiconductors (see the Introduction to this volume), and among them the microwave methods occupy a special place because they provide a rich trove of information about the properties of nonequilibrium carriers and the condensation process. The effectiveness of microwave methods has been demonstrated in studies of exciton condensation in germanium. There have been studies of the kinetics of both microwave conductivity and breakdown of excitons in the presence of electron-hole droplets (EHD)[1-5],* cyclotron resonance of the free carriers surrounding an EHD [6], and Alfvén size resonances in EHD [7, 8].

We wish to extend the analysis of the data obtained elsewhere [1-4] in investigations of microwave conductivity and microwave breakdown of excitons in germanium in order to show what information can be extracted concerning exciton condensation. It should be noted that photoconductivity studies of semiconductors in the microwave region have several advantages when compared with constant-current photoconductivity measurements. The microwave approach eliminates contact effects (the photoeffect and carrier injection at electrodes). In addition, the high sensitivity of the microwave methods enables one to study the behavior of nonequilibrium carriers over a wide range of densities in samples of diverse shapes and sizes.

When pulsed optical excitation is combined with microwave conductivity the result is a simple and clear method for studying the evolution of a system of nonequilibrium carriers, the process of exciton formation and condensation, the diffusion of free and bound carriers away from the excitation region, and for determining their density. By using microwave breakdown of excitons in conjunction with microwave conductivity one can obtain extensive information about the system composed of free carriers (FC), free excitons (FE), and electron-hole

---

* Previous articles in this review treat these works in greater detail.

droplets. We shall now analyze more thoroughly the microwave methods used to investigate exciton condensation.

Consider the kinetics of an FC−FE−EHD system in a semiconductor after pulsed excitation of carriers (as by a laser pulse) in the presence of a microwave-frequency electric field. The field is assumed to consist of a weak field E_1 used to record the microwave conductivity of the sample, and a much stronger field E_2 which greatly heats the free carriers and causes the excitons to be ionized by collisions. The frequencies of the two fields ω_1 and ω_2 may be either identical or different. A number of studies have preferred to use the latter case in which the frequency difference $\omega_1 - \omega_2$ is much greater than the pass band of the microwave spectrometer. We shall assume that the carriers are uniformly distributed throughout the sample. This situation can be realized either with surface generation of carriers by one-photon excitation of samples with dimensions comparable to the characteristic carrier diffusion length, or with bulk generation of carriers as in multiphoton excitation. The kinetics of the FC−FE−EHD system are described by the following rate equations for the free carrier density n_e, the free exciton density n, and the density of carriers in a droplet N:

$$dn_e/dt = \alpha n n_e - \beta n_e^2 - (\Gamma_e + \tau_e^{-1})n_e + w_A N_d N_\triangle;$$
$$dn/dt = -\alpha n n_e + \beta n_e^2 - \Gamma(n - n_\mathrm{T}) - n/\tau_{ex};$$
$$dN/dt = \Gamma(n - n_\mathrm{T}) + \Gamma_e n_e - N/\tau_0. \tag{1}$$

Here the term $\alpha n n_e$ describes the collision ionization of the exciton gas in the electric field E_2 [$\alpha = \alpha(E_2)$ is the collision ionization coefficient], βn_e^2 describes the binding of carriers into excitons (β is the coupling coefficient), the capture of carriers by EHD's and lattice defects is given by the terms $\Gamma_e n_e$ and $\tau_e^{-1} n_e$ ($\Gamma_e = 4\pi R^2 N_d v$ is the rate at which carriers are trapped by drops of radius R, where N_d is the density of drops in the sample, v is the characteristic time for the capture of carriers by defects), $w_A N_d N_\triangle$ accounts for the ejection of carriers from an EHD through Auger recombination (w_A is the probability of the Auger process, $N_\triangle = 4\pi R^2 n_0 \Delta$ is the number of carriers in a surface layer of thickness Δ from which Auger carriers are ejected*), $\Gamma(n - n_\mathrm{T})$ accounts for exciton capture by, and evaporation from, drops ($\Gamma = 4\pi R^2 N_d v_\mathrm{T}$ is the rate for capturing excitons having thermal velocity v_T), n_T is the equilibrium exciton density at temperature T, $n\tau_{ex}^{-1}$ describes the recombination of free excitons, and $N\tau_0^{-1}$ gives the volume recombination of carriers inside an EHD. Thermal dissociation of excitons and carrier evaporation from drops are not included in these equations because they are negligible effects at low temperatures. Calculations indicate that Auger processes are the main source of free carriers in a system with T < 2°K in the absence of a microwave field. While the application of a very strong microwave field can heat both free carriers and EHD's, we have included just the first process in the rate equations (i.e., the collision ionization of excitons in the microwave electric field) because it has been studied in detail experimentally whereas microwave heating of drops has yet to be reliably observed.†

The solution to Eq. (1) has been examined in detail in [2, 3] (without, however, including the terms describing carrier ejection from a drop through Auger processes; see the first

---

*We show below that Auger carriers are effectively ejected only from a surface layer rather than from the interior of a drop as has been suggested in a number of papers [6, 9, 10].

† In some experiments in our laboratory it appears that heating of drops has been observed, but we still have to understand the conditions under which this effect is most easily produced. In particular, one would expect that drop heating should be more effective in a microwave magnetic field since the electric field of the microwave penetrates very little into the drop because of its high conductivity. Under certain conditions it is possible to heat drops indirectly by the electric field because the drop is bombarded by hot carriers formed in collision ionization of excitons.

article in this volume). Therefore, we shall present here only the basic, most important results of the solutions to these equations and an analysis of the experimental data obtained in [2-5] based on these solutions.

In the absence of collision ionization of excitons, i.e., when the microwave electric field strength E is lower than the threshold field E_c, the densities of free carriers and free excitons some time after the exciting laser pulse, when drops are formed, follow these quasistationary solutions:

$$n_e(t) = \frac{w_A N_d N_\Delta(t)}{\Gamma_e(t) + \tau_e^{-1}} ; \qquad (2)$$

$$n(t) = \frac{\Gamma(t) n_T}{\Gamma(t) + \tau_{ex}^{-1}} . \qquad (3)$$

It is reasonable to assume that the number of drops remains constant over a considerable period of time, possibly with the exception of the end of the process when the drop radius is very small. In order to explain the observed kinetics of microwave conductivity in Ge at 1.6°K (see the preceding article of this volume) we must assume that in these samples the rate of carrier capture by lattice defects is greater than that of EHD's ($\tau_e^{-1} \gg \Gamma_e$). Thus, the free carrier density follows the kinetics of the Auger carriers ejected from the drop surface layer with a time constant 1.5 times greater than the "bulk" lifetime of a drop τ_0:

$$n_e(t) = w_A \tau_e N_d N_\Delta \sim R^2(t) = \bar{R} \exp(-\tfrac{2}{3}t/\tau_0). \qquad (4)$$

Note that the assumption that $\tau_e^{-1} \gg \Gamma_e$ is completely valid for cold carriers (i.e., in a weak electric field), but can become invalid for hot carriers which are heated by the microwave electric field [Γ_e increases with the carrier velocity; see Eq. (1)]. In fact, as we shall see below, when the exciton gas in the presence of EHD's breaks down in the microwave field, the drops become effective traps for hot carriers; for these carriers the condition $\Gamma_e(E) \gtrsim \tau_e^{-1}$ is satisfied, and this determines the breakdown characteristics.

Equation (4) provides a good description of the observed kinetics of microwave conductivity in Ge when laser excitation is used. This is reflected in the fact that the observed characteristic time for the conductivity kinetics, which is about 50 μsec at 1.6°K [2], is a factor of 1.5 greater than the EHD lifetime $\tau_0 \simeq 35\,\mu$sec as measured in independent microwave breakdown experiments (see below) and from EHD recombination radiation experiments [5]. Such good agreement between the values of τ_0 obtained from studies of EHD under different physical conditions is a strong argument in support of the assumption that Auger carriers are ejected from a surface layer of the drop rather than from the bulk of the drop as suggested elsewhere [6, 9, 10]. The mean free path l_A of hot carriers formed in an Auger process turns out to be smaller than the drop radius R. By comparing the density of free carriers obtained from the kinetics of microwave conductivity with the predictions of Eq. (3), we can obtain the effective thickness Δ of the surface layer from which the Auger carriers are ejected, and then we may estimate the effective collision frequency ν_A of carriers in the drop. For germanium at 1.6°K we find that $\Delta \simeq 10^{-5}$ cm and $\nu_A \simeq 10^{12}$ sec^{-1}. We used the following values in our calculations: $w_A = 0.8 \cdot 10^4$ sec^{-1} (assuming that $w_A = \tfrac{1}{2}cn_0^2$, this quantity can be obtained from the Auger recombination coefficient $c = 4 \cdot 10^{-31}$ cm$^6 \cdot$ sec^{-1} as measured in [11] from oscillations in the luminescence intensity from EHD's in a magnetic field); $N_d = 5 \cdot 10^4$ cm^{-3} (measured in light-scattering experiments in EHD's [12]); $\tau_e = 10^{-6}$ sec, $\bar{R} = 10^{-3}$ cm (measured in experiments on the microwave breakdown of excitons [2]; this value of $\bar{R}$ also corresponds to measured values in light-scattering experiments [12, 13]); $v_A = 10^7$ cm/sec (the rate at which Auger carriers are ejected).

Equation (3) suggests that the density of free excitons in equilibrium with an EHD must be essentially constant over an extended period following the drop formation because the cross section for exciton capture by a drop $\Gamma(t)$ can be much larger than the rate at which free excitons recombine ($\Gamma \gg \tau_{ex}^{-1}$). This result corresponds to the usual physical situation with ordinary liquids; the pressure of the saturated vapor above the liquid is constant. Experiments to determine the exciton gas density in the presence of drops in Ge at 1.6°K (by measuring the density of free carriers formed by microwave breakdown of the exciton gas [see below] have confirmed this result (see [4] and the second article of this review). Equation (1) and the value of n_T can be used to find the work function ψ for carriers escaping a drop. Although detailed measurements of the free exciton density n_T with the indicated method have not been completed for a range of temperatures, the measurements at 1.6°K [4] give $n_T \simeq 10^{12}$ cm^{-3}, in agreement with $\psi \simeq 2$ meV as measured in recombination radiation experiments (see [9]). Studies of the kinetics of n by this method at the end of the EHD decay process, when their radius and exciton capture cross section are both small [$R(t) < 1\ \mu$m, $\Gamma(t) < \tau_{ex}^{-1}$], result in a free exciton lifetime of $\tau_{ex} = 8 \cdot 10^{-6}$ sec which agrees well with direct observations of the intensity of the exciton luminescence line.

Let us now consider the microwave breakdown of the exciton gas in the presence of EHD's as a result of collision ionization of excitons by the microwave electric field. The theory for this effect was developed in [2] (see the first article of this volume). We shall derive a number of important results from this theory.

The threshold for the breakdown effect, which is defined as the appearance of a large number of free carriers, is very dependent on the duration τ of the microwave pulse. When $\tau < \tilde{\Gamma}_e^{-1}$ ($\tilde{\Gamma} = \Gamma_e + \tau_e^{-1}$) this threshold criterion takes the form

$$a\,(E_c)n_T\tau = 1, \tag{5}$$

where E_c is the threshold (critical) field at which breakdown begins. This criterion has a very simple physical interpretation: Every electron must undergo at least one ionizing collision during the electric field pulse. When $\tau > \tilde{\Gamma}_e^{-1}$ the criterion for breakdown of the exciton gas is

$$a\,(E_c)n_T\tau_{ex} = \xi, \tag{6}$$

where

$$\xi = \frac{v_E}{v_T} + \frac{\tau_{ex}}{\tau_e} + 2\sqrt{\frac{v_E}{v_T}\frac{\tau_{ex}}{\tau_e}}\cosh\left(\frac{2}{3}\frac{t-t_0}{\tau_0}\right),$$

$$t_0 = {}^3\!/_2\tau_0\ln\left(\sqrt{\frac{v_E}{v_T}\tau_e\tau_{ex}}\,\Gamma_0\right),\quad \Gamma_0 = \Gamma\,(t=0),$$

v_T is the thermal velocity of the free carriers, and v_E is their velocity in the field E.

It follows from Eqs. (5) and (6) that the threshold field E_c begins to increase when the pulse duration decreases to $\tau = \tilde{\Gamma}_e^{-1}$. Thus, measurements of the dependence of the exciton breakdown threshold on the duration of the microwave pulse enable one to find the cross section for carrier trapping by EHD's and defects and to compute both the radius and number of drops.

It also follows from Eq. (6) that the threshold field E_c is the smallest when $t = t_0$. This minimum threshold as a function of delay time is related physically to time-dependent changes in the drop radius (since during recombination of carriers their density remains constant); the drop is an effective trapping center for carriers. Measurements of t_0 as a function of I, the intensity of the exciting laser pulse, which determines the initial drop radius (since $\Gamma_0 \sim \overline{R}^2$, which gives $\Gamma_0 \sim I^{2/3}$ for one-photon excitation, for example), are a good method for measuring the drop lifetime τ_0. For germanium such measurements give $\tau_0 = 35$ μsec at 1.6°K for both one-photon [3] and two-photon [5] laser excitation. As noted above, this result is in very good agreement with the data from microwave conductivity kinetics and recombination radiation measurements.

When the microwave field strength is much greater than the threshold value ($E \gg E_c$) a sharp peak is observed in the sample conductivity. Clearly the amplitude and shape of this peak must depend on the density of the exciton gas, the ionizing collision cross section, and the capture of carriers by drops and defects. In a strong microwave field and with breakdown of the exciton gas well developed, the solution to Eq. (1) for the maximum concentration of free carriers is

$$n_{e\,\max} = \tilde{n}_e + n\left[1 - \frac{\widetilde{\Gamma}_e}{a\tilde{n}}\left(1 + \ln\frac{a\tilde{n}}{\widetilde{\Gamma}_e}\right)\right],\tag{7}$$

and the characteristic times for buildup and decay of the density of free carriers (the leading and trailing edges of the breakdown peak) are respectively

$$\Delta t_1 = (an_\tau)^{-1}, \quad \Delta t_2 = \widetilde{\Gamma}_e^{-1}.\tag{8}$$

Here $\tilde{n}_e = n_e(t_0)$ and $\tilde{n} = n(t_0)$ are the initial densities of free carriers and free excitons at the moment t_0 the breakdown microwave field is applied.

At low temperatures it is observed that the edges of the breakdown peak in the microwave conductivity are $\Delta t_1 \simeq \Delta t_2 \simeq 10^{-7}$ sec, so that from Eq. (7) we find $n_{e\,\max} \simeq \tilde{n} = n_\tau$. This latter result indicates that the microwave electric field is sufficiently strong (the breakdown threshold E at 1.6°K is about 10 V/cm [1]) to totally break down the exciton gas. Thus, a measurement of the amplitude of the breakdown peak provides a direct measurement of the exciton density n_T in equilibrium with the electron-hole droplet. This approach to n_T was used in [3] with the double-frequency microwave technique (wherein the photoconductivity is measured at $\omega_1 = 10$ GHz while the exciton gas is broken down at a frequency ω_2 in the range 0.3–0.9 GHz), and its effectiveness was demonstrated down to very low temperatures (T = 1.3°K). The measured values of n_T are in good agreement with the well-known equation [14]

$$n_\tau = g\left(\frac{MkT}{2\pi\hbar}\right)^{3/2}\exp\left(-\frac{\psi}{kT}\right),\tag{9}$$

where g = 16 is the degeneracy of the exciton ground state, M = 0.335 is its effective mass, and $\psi \simeq 2$ meV is the work function for excitons leaving a drop [15]. Typical values of n_T are about 10^{12} cm^{-3} at 1.5°K. Direct measurements of such small equilibrium exciton densities by means of luminescence intensity from the exciton line are very difficult. The highly sensitive microwave spectrometers can easily record free carrier densities of the order of 10^{10} cm^{-3}, and when combined with microwave breakdown of the exciton gas one can measure n_T even at very low temperatures (T < 1.3°K). Such measurements are of interest for determining the liquid–gas phase diagram for excitons and for studying the properties of EHD's at very low temperatures.

The shape of the breakdown peak is important for understanding the dynamics of free carriers in the exciton gas in the presence of EHD's. In particular, Eq. (7) indicates that the shape allows one to determine the cross section for collision ionization of excitons and the cross section for capture of free carriers by drops and defects.

The abrupt character of the exciton gas breakdown observed at low temperatures (T < 2.5°K [2–5]) is due to the fact that the hot free carriers formed by destroying excitons are

TABLE 1. Parameters of the Free Carrier – Free Exciton – EHD System in Ge, Measured with Microwave Conductivity and Microwave Breakdown of the Exciton Gas at 1.6°K

Lifetimes:	
Free Carriers	$\tau_e \simeq 1 \; \mu sec$
Free Excitons	$\tau_{ex} \simeq 8 \; \mu sec$
Electron-hole drops	$\tau_0 \simeq 35 \; \mu sec$
Densities in Thermal Equilibrium	
Free Carriers	$n_{eT} \simeq 10^{10} \; cm^{-3}$
Free Excitons	$n_T \simeq 10^{12} \; cm^{-3}$
Electron-hole pairs in EHD	$n_0 \simeq 2 \cdot 10^{17} \; cm^{-3}$
Drops in sample	$N_d \simeq 5 \cdot 10^4 \; cm^{-3}$
Initial drop radius ($\bar{n}_e = 10^{15} \; cm^{-3}$)	$\bar{R} \simeq 10^{-3} \; cm$
Density of Auger carriers ejected from EHD	$\bar{n}_{eA} \simeq 10^{10} \; cm^{-3}$
Effective thickness of layer from which Auger carriers are ejected	$\Delta \simeq 10^{-5} \; cm$
Effective collision frequency of carriers in EHD	
for Auger carriers	$\nu_A \simeq 10^{12} \; sec^{-1}$
for thermal carriers	$\nu_T \simeq 10^{11} \; sec^{-1}$

captured by drops and defects much faster than the carriers can bind into excitons and faster than excitons can evaporate from the drops. Under certain conditions, in particular at higher temperatures, the rate at which excitons evaporate from drops increases and it is possible to have a steady-state regime for breakdown of the exciton gas during the period the microwave field is on.

An interesting effect is observed in the peak breakdown mode [2, 3] when two successive microwave pulses are employed. When the interval between the pulses is shorter than about 10 μsec the second pulse does not cause breakdown. However, when $\Delta t > 10 \; \mu$sec the second pulse causes breakdown just as easily as the first. This indicates that the time required to reach the equilibrium density of excitons at that temperature because of evaporation from the drops is about 10 μsec, in good agreement with the cross section Γ for exciton evaporation obtained from Eq. (1).

We wish to draw attention to an interesting circumstance connected with the abrupt mode of exciton breakdown. After the complete breakdown of the exciton gas the electron-hole drop finds itself in an "exciton vacuum" for a period of about 10 μsec. In this situation one might expect some effects which are characteristic of a common liquid under similar conditions (for example, boiling).

The above analysis demonstrates that the kinetics of microwave conductivity when combined with microwave breakdown of the exciton gas serve as a rich source of information concerning exciton condensation and the system composed of free carriers, free excitons, and electron-hole droplets. To illustrate the point, we present in Table 1 the more important parameters for such a system in germanium as measured at the present time in our laboratory.

LITERATURE CITED

1. A. A. Manenkov, V. A. Milyaev, G. N. Mikhailova, and S. P. Smolin, Pis'ma Zh. Éksp. Teor. Fiz., 16:454 (1972).
2. L. V. Keldysh, A. A. Manenkov, V. A. Milyaev, and G. N. Mikhailova, Zh. Éksp. Teor. Fiz., 66:2178 (1974).
3. L. V. Keldysh, A. A. Manenkov, V. A. Miljaev, and G. N. Mikhailova, Proceedings of the Twelfth International Conference on the Physics of Semiconductors, Stuttgart, 1974, p. 76.

4. A. A. Manenkov, V. A. Milyaev, G. N. Mikhailova, V. A. Sanina, and A. S. Seferov, Zh. Éksp. Teor. Fiz., 70:695 (1976).
5. B. V. Zubov, A. A. Manenkov, V. A. Milyaev, G. N. Mikhailova, T. M. Murina, and A. M. Prokhorov, Fiz. Tverd. Tela, 18:2024 (1976).
6. J. C. Hensel, T. G. Phillips, and T. M. Rice, Phys. Rev. Lett., 30:227 (1973).
7. R. S. Markiewicz, J. P. Wolfe, and C. D. Jeffries, Phys. Rev. Lett., 32:1357 (1974); 34:59(E) (1975).
8. J. P. Wolfe, R. S. Markiewicz, C. Kittel, and C. D. Jeffries, Phys. Rev. Lett., 34:275 (1975).
9. C. D. Jeffries, Science, 189:955 (1975).
10. P. S. Gladkov, B. G. Zhurkin, and N. A. Penin, Fiz. Tekh. Poluprovodn. 6:1919 (1972).
11. K. Betzler, B. G. Zhurkin, and A. Karuzskii, Solid State Commun., 17:577 (1975).
12. N. N. Sibel'din, V. S. Bagaev, N. A. Penin, and V. A. Tsvetkov, Fiz. Tverd. Tela, 15:177 (1973).
13. V. S. Bagaev, N. A. Penin, N. N. Sibel'din, and V. A. Tsvetkov, Fiz. Tverd. Tela, 15:3269 (1973).
14. L. V. Keldysh, in: Excitons in Semiconductors [in Russian], Nauka, Moscow (1971), p. 5.
15. T. K. Lo, B. J. Feldman, R. M. Westervelt, J. L. Staehli, C. D. Jeffries, and E. E. Haller, Preprint N UCB-34P20-167 (1974).

A HELIUM-3 REFRIGERATOR FOR STUDYING EXCITONS IN Ge AT TEMPERATURES BELOW 1°K

S.I. Valyanskii, V.A. Milyaev, G.N. Mikhailova, and A.B. Fradkov

The construction and test results for a He^3 refrigerator are presented. The refrigerator is used to study microwave conductivity and recombination radiation in semiconductors with the use of laser radiation. The cooling power of the refrigerator is 10^{-2} W at 0.5°K. The problem of sample warming by the laser radiation is examined.

There are a number of methods used for producing temperatures lower than 1°K; pumping over liquid helium-4 and helium-3, adiabatic demagnetization of paramagnetic salts, and dissolving He^3 in He^4. Adiabatic demagnetization makes it possible to reach very low temperatures (~0.005°K), but the apparatus has a very low capacity for cooling and one must operate in a reheating mode. Recently refrigerators based on dissolving He^3 in He^4 have become quite popular [1]. However, this method requires very complex equipment and at present most experiments conducted with such refrigerators have been limited to the cryogenics problems themselves.

The lowest temperature achieved by pumping on He^4 is about 0.7°K. Much lower temperatures can be reached if one uses He^3 as the cooling agent. This is because the boiling point of He^3 is 3.2°K and its vapor pressure is higher than that of liquid He^4 at any temperature. If the pressure above the liquid is reduced to some value, the temperature of He^3 at that pressure will always be lower than that of He^4 at the same pressure. Moreover, He^3 does not become superfluid until the temperature is lowered to at least 0.003°K, and it does not contain any film currents transporting both liquid and heat. It can be pumped to much lower pressures than He^4. Thus, He^3 is attractive as a cryogen for producing temperatures below 1°K.

Helium-3 is a stable but rare isotope of helium. The He^3 content of a helium atmosphere is $1.2 \cdot 10^{-6}$ [2], but the fraction by volume of the earth's atmosphere is less than 10^{-11}. Although it is possible in practice to obtain pure He^3 by successive enrichment through various processes, this approach is very tedious. Suffice it to say that in order to obtain 1 liter of gaseous He^3 at standard condition one would have to process 1000 m^3 of helium obtained from natural sources.

Advances in nuclear physics have led to a new method for producing pure He^3. The method involves the production of tritium by irradiating lithium with neutrons: Li^6 + n → He^4 + H^3. The helium-3 is a by-product of the β-decay of tritium (H^3 → He^3 + e) with a half-

* This work was completed in 1973.

67

life of 11.6 years. The primary difficulty associated with this method is the separation of the He³ from the radioactive tritium.

In spite of the fact that He³ is now cheaper and more readily available, one must exerise many precautions when working with it in order to reduce losses and contamination of the gas.

The first He³ refrigerator in the Soviet Union was constructed by Peshkov and Zinov'eva [3]. We have constructed a similar system [4], a diagram of which is shown in Fig. 1. The system consists of a cryostat, pumps for He³ and He⁴, storage tanks for He³, manometers, and associated tubing (Fig. 2). The liquid He⁴ cryostat (1) contains the evaporative He³ bath (2) which is surrounded by a vacuum jacket (3) for thermal isolation. The He⁴ is reduced to about 1.3°K (p = 0.6 mm Hg) by the VN-6 pump. The He³ is stored in 30-liter tanks (4) at a pressure of 600 mm Hg. The He³ leaves the tanks (4), is purified by an adsorbing carbon trap (9) at nitrogen temperature, goes to the heat exchanger (5) which is immersed in 1.3°K He⁴, and is then throttled in the capillary tube (6) and condensed into the evaporative bath (2).

The condensed He³ (now at a temperature of 1.3°K) is pumped by a series of pumps: a TsVL-100 oil diffusion pump, a DRN-50 mercury diffusion pump, and an NVG-2 sealed fore-pump. The He³ is thus cooled to an equilibrium temperature which is determined by the heat load and the pumping rate ($T_{min} = 0.3°K$, $p \sim 0.002$ mm Hg). The evaporated gaseous He³ exhausted by the NVG-2 pump can be directed again to the heat exchanger (5) to be recooled and sent to the throttle (6). In this case the system operates in a continuous cycle; the He³ level

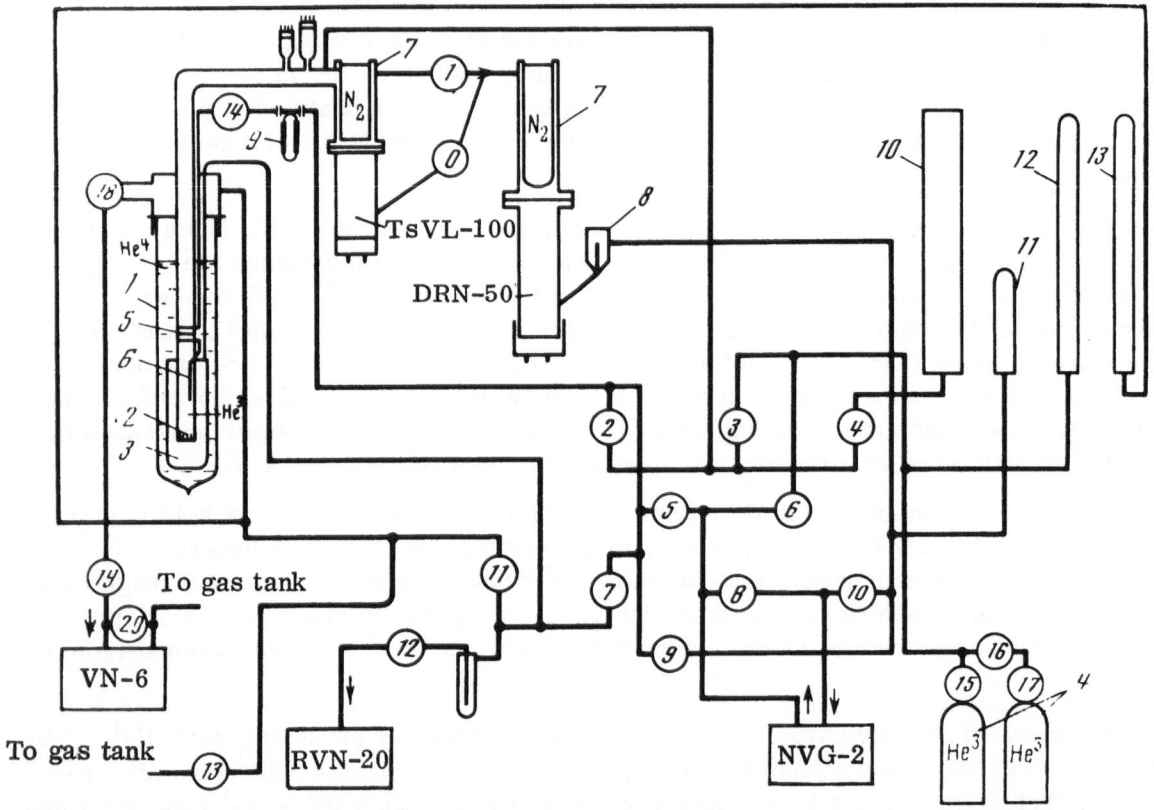

Fig. 1. Apparatus for producing temperatures below 1°K with He³. 1, He⁴ cryostat; 2, evaporative bath for He³; 3, vacuum jacket; 4, He³ storage tanks; 5, heat exchanger; 6, throttle capillary; 7, nitrogen traps for pumps; 8, mercury vapor trap; 9, adsorptive carbon trap cooled by liquid nitrogen; 10, McLeod manometer; 11-13, mercury manometers. Numbers in circles are values.

Fig. 2. Photograph of He^3 system.

in the evaporative bath is constant and the time available for experimentation is limited only by the He^4 present in the outer Dewar (1).

The auxiliary pump RVN-20 pumps down the tubing before the He^3 is admitted. The system is equipped with four manometers for pressure measurements: one at the exhaust of the DRN-50 pump (11), one at the He^3 storage tanks (12), and one on the liquid He^4 (13). The fourth, a McLeod manometer (10) [5], measures the pressure of the saturated He^3 vapor and, hence, the temperature in the evaporative bath.

The study of high-density excitons in semiconductors at temperatures below 1°K necessitates the development of a new cryostat-refrigerator, one which will allow complex optical and microwave investigations at such temperatures. All the experimental work done thus far on exciton condensation in Ge has been conducted above 1°K. This is not surprising in view of the experimental problems; one must simultaneously ensure effective optical excitation of the sample and still cool it to the lowest possible temperature. The study of excitons at high densities and temperatures below 1°K is of great theoretical interest [6] because one expects some new phenomena in the exciton system at very low temperatures. We have studied microwave conductivity and luminescence from Ge with laser excitation at temperatures in the range from 0.5-1.5°K using He^3 as a cryogen. Although the literature contains descriptions of a number of He^3 cryostats for EPR spectroscopy [7-12], these systems do not permit optical excitation of the sample. Therefore it was necessary to design a new optical cryostat-refrigerator combination with increased cooling power and easy access to the experimental chamber for changing the sample.

Faced with these requirements, we have constructed a He^3 refrigerator for microwave measurements with laser excitation [13]. An overview of the refrigerator is shown in Fig. 3 and a detail of the lower portion is shown in Fig. 4. It consists of either a glass or metal

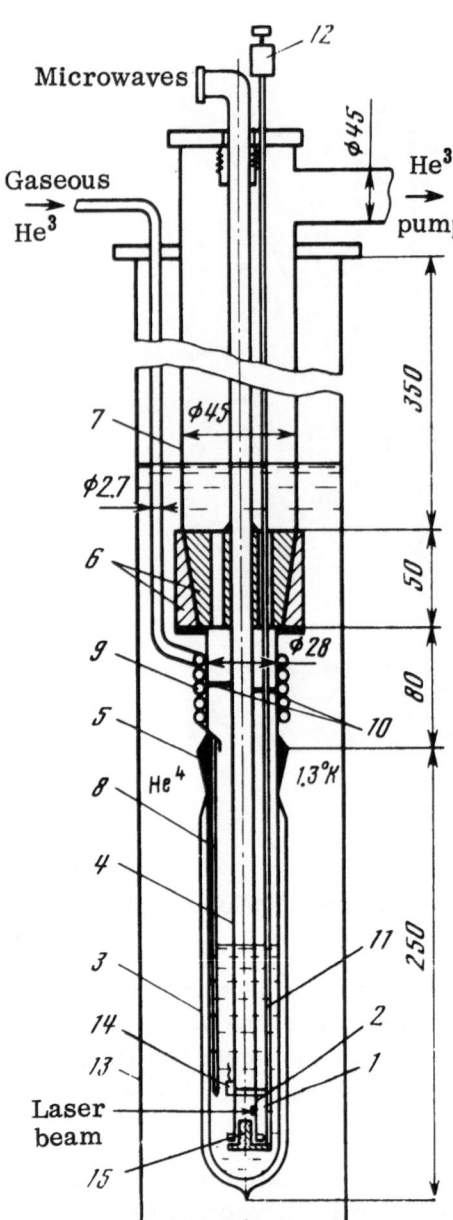

Fig. 3. Schematic of He³ refrigerator for microwave measurements with laser excitation. 1, Ge sample; 2, lower waveguide section; 3, He⁴ Dewar; 4, 36-GHz waveguide (7.2 × 3.4 mm); 5, copper-glass joint; 6, joint; 7, He³ pump tube; 8, throttle capillary; 9, heat exchanger; 10, radiation shield; 11, rod for moving piston; 12, bellows unit for controlling rod; 13, He⁴ cryostat; 14, carbon thermometer; 15, piston.

Dewar with quartz windows for holding about 3 liters of He⁴; the He³ chamber is placed inside. The evaporative bath for He³ is a glass Dewar (3) of about 100 cm³ capacity. A glass He³ Dewar is a simple solution to two problems. First, it permits optical radiation to enter the chamber without significant radiation loss. Second, it provides vacuum isolation for the He³. The Dewar is covered with aluminized Dacron film to guard against thermal radiation. The upper part of the Dewar (3) has a glass-to-copper joint (5) where the Dewar is attached to the He³ pump tube (7). This tube (7) is made of 0.14-mm-thick stainless steel. The radiative heat load along the pump tube from the warm cryostat surfaces to the He³ bath is removed by the radiation shields (10) made of polished copper foil.

The heat exchanger (9) and capillary (8) serve to condense the He³ in the cycling mode. The heat exchanger is a stainless steel coil of diameter 2.7 mm, 0.2 mm thick, and about 2.5 m long. The throttle capillary is German silver tubing with an inside diameter of 0.2 mm and 250 mm long.

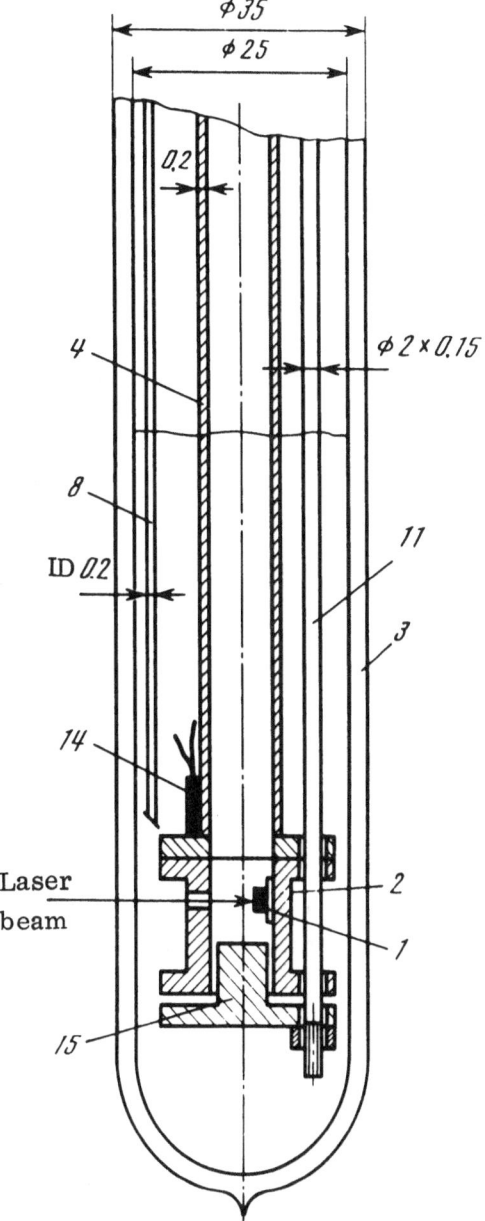

Fig. 4. Detail of the lower section of the He³ refrigerator. Same notation as in Fig. 3.

The apparatus described here is used for studying the microwave conductivity of Ge at 36 GHz (λ = 8 mm). The sample (1) is placed in a 3.4 × 7.2 mm² section of waveguide (2) terminated by the piston (15) and cooled by the evaporative He³ bath (3). The microwave circuit is tuned by means of the rod (11) soldered to the bellows unit (12) on the upper flange. The waveguide is hermetically sealed by a quartz disc glued into the exit end of the waveguide with PR-epoxy [14]. The waveguide with the sample is inserted into the He³ chamber from above and sealed with a flanged joint. This simplifies disassembly and increases reliability because it is not necessary to unsolder the cryostat in order to change samples. The waveguide and rod are made of thin-wall stainless steel tubing (0.2 mm thick) and the waveguide is about 900 mm long.

In order to reduce the heat into the He³ bath from the waveguide and control rod, both of which have parts at room temperature and which communicate with the He³ bath, the waveguide and rod are thermally connected to the He⁴ bath by means of a massive copper joint (6) [15].

The inner portion of the joint is a core with eight openings 8 mm in diameter for pumping on the He^3. It is soldered to the waveguide with PSR-45 solder. The outer part of the joint, the connecting sleeve, is bathed in He^4 liquid. The movable rod is thermally connected to the joint with a piece of flexible copper wire with a cross section of 4 mm^2. The waveguide is soldered to the upper flange of the He^3 chamber through a sylphon bellows to compensate for temperature variations which could cause the joint to loosen.

The Ge sample is greased to the narrow wall of the waveguide with Apiezon grease. Since He^3 has a low thermal conductivity it is possible to find the liquid layered in the evaporative bath near parts having different temperatures. We reduce this effect by bathing a copper part on the waveguide in He^3 and fitting it with a large-area radiator to average out the bath temperature. A carbon resistor (14) [16] is attached at this point to measure the temperature. The carbon thermometer is calibrated from the pressure of the saturated He^3 vapor as measured by the McLeod manometer and converted to the T_{62} scale. The thermometer resistance is measured potentiometrically with a current of about 1 μA. The leads to the resistor are made of 0.07-mm-diameter manganin.

The Ge sample is optically excited by means of a pulsed Nd^{3+}YAG laser ($\lambda = 1.06 \mu m$). The maximum pulse energy is 10^{-4} J and the pulse repetition rate is 100 Hz. The radiation strikes the sample after passing through a window in the He^3 Dewar and a 1-mm-diameter opening on the microwave guide. About 50% of the light is lost in the process.

A visual examination of the He^3 showed that the liquid did not boil vigorously when the laser radiation passed through it. This is because He^3 has a low absorption coefficient for 1.06 μm radiation.

Our measurements indicate that the cooling power of the refrigerator is 10^{-2} W at 0.5°K and that the lowest temperature in the cycle mode is 0.45°K. In this mode one loses about 1 liter of He^4 every hour. We have evaluated the primary sources of heat input to the He^3 bath of our refrigerator. They are

a. heat flowing along the Dewar walls, the waveguide, the control rod, and the electrical leads;
b. heat from radiation from the He^4 bath;
c. Joule heating in the carbon thermometer and its leads.

The heat flux Q_T across a solid body of cross-sectional area A because of a thermal gradient $\partial T/\partial x$ is $Q_T = \lambda(T) A \partial T/\partial x$, where $\lambda(T)$ is the thermal conductivity. When the flux is small and the temperature gradient is low we may assume that $\partial T/\partial x$ is a constant and we can use the mean integral value $\bar{\lambda}$ for $\lambda(T)$. Then $\bar{Q}_T = \bar{\lambda}(A/l)\Delta T$, where ΔT is the temperature difference and l is the length.

We shall compute the flux Q_T along the Dewar wall from the upper end, which is at 1.3°K, to the liquid He^3 at T = 0.5°K. The inner vessel of the Dewar has the following dimensions: diameter 25 mm, l = 200 mm (to the He^3 level), and a wall thickness of about 1 mm. For glass we have λ = 0.0007 W/cm-deg [17], and ΔT = 0.8°. Thus, $Q_{T_1} \approx 2.1 \cdot 10^{-5}$ W.

Now let us calculate the heat flux Q_{T_2} carried by the walls of the waveguide (7.2 × 3.4 mm cross section, 0.2 mm thick, stainless steel). Assuming that the waveguide joint is held at the He^4 temperature we have ΔT = 0.8° and l = 300 mm. The thermal conductivity of stainless steel is $\bar{\lambda}$ = 0.001 W/cm-deg [18]. Thus $Q_{T_3} \approx 10^{-6}$ W. Analogously, for the rod (diameter 2 mm, thickness 0.15 mm, l = 300 mm) we have $Q_{T_2} \approx 2.8 \cdot 10^{-7}$ W. The heat flux due to electrical conduction (2 wires, diameter 0.07, $l \simeq 30$ cm, manganin, λ = 0.006 W/cm-deg [17]) is $Q_{T_4} \approx 10^{-7}$ W. Thus the total heat flux due to thermal conduction is $Q_T \approx 2.2 \cdot 10^{-5}$ W.

The thermal radiation Q_r from the outer surface of the He^3 Dewar, which is covered with aluminized foil at 1.3°K, to the copper waveguide inside the He^3 bath (at 0.5°K) can be calculated from the Stefan–Boltzmann equation by taking the emissivity of the aluminum foil to be $\varepsilon_1 = 0.011$ [17] and that of the copper to be $\varepsilon_2 = 0.084$ [17]. When $\varepsilon_1 \simeq \varepsilon_2 \ll 1$ [17]

$$Q_r = {}^1\!/_2 \varepsilon \sigma A (T_1^4 - T_2^4).$$

Here $\sigma = 5.67 \cdot 10^{12}$ W/(cm-deg^2)^2 is the Stefan–Boltzmann constant, A is the area of the radiating surface, and T_1 and T_2 are the temperatures of the radiating and absorbing surfaces. We find that $Q_r \approx 10^{-9}$ W. Thus, when the He^3 bath is not receiving radiation from either room-temperature or liquid-nitrogen-temperature objects, the heat flux Q_r is negligible.

The Joule heat liberated in the carbon resistance thermometer and the electrical leads is $Q_J = I^2(R_{th} + R_l)$; $R_T = 10$ kΩ when $T = 0.5$°K, $R_l \simeq 30$ kΩ. The current through the thermometer is 1 μA. Thus $Q_J = 10^{-8}$ W. The total heat flux to the He^3 bath is therefore equal to

$$Q = Q_T + Q_r + Q_J \simeq 2.2 \cdot 10^{-5} \text{ W}.$$

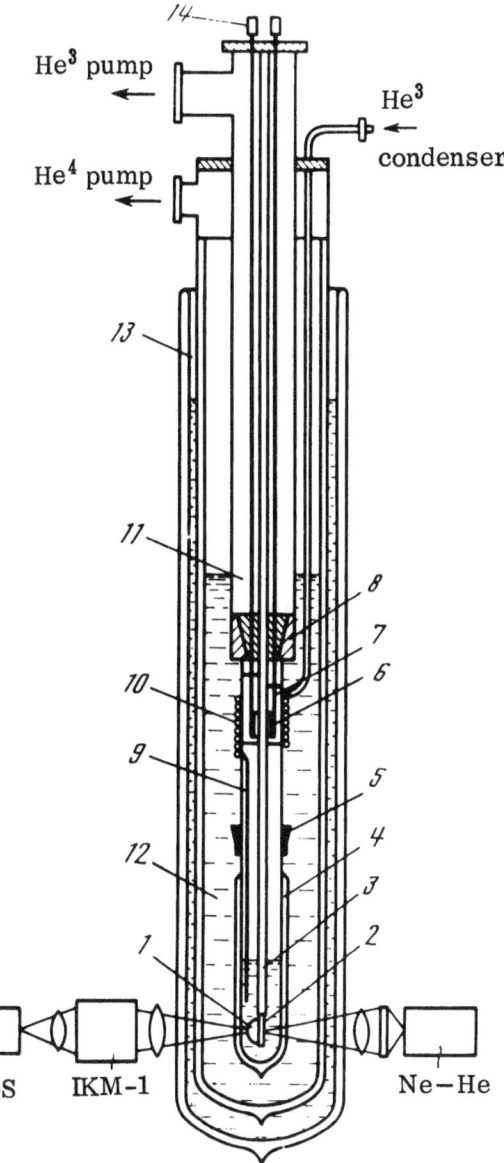

Fig. 5. Drawing of the He^3 refrigerator for studying recombination radiation in Ge at temperatures between 0.5°K and 1.5°K. 1) Ge sample; 2) copper holder; 3) suspension rod; 4) He^3 Dewar; 5) copper–glass joint; 6) aligning apparatus; 7) radiation screen; 8) joint; 9) throttle capillary; 10) heat exchanger; 11) He^3 pumping tube; 12) He^4 Dewar; 13) nitrogen Dewar; 14) bellows unit. IKM-1 is a monochromator.

The cooling power of the system W = 10^{-2} W at 0.5°K agrees with the calculated values for this quantity based on the equation W = qV, where the heat of evaporation of He^3 is q = 30 J/mole at 0.5°K [19] and V is the rate at which He^3 is circulated (V = $3 \cdot 10^{-4}$ mole/sec). The rate V, which determines the refrigerator's cooling power, depends on the pumping capacity, the tube diameters, and the efficiency of the heat exchanger. Detailed calculations for the evaporation refrigerator using He^3 are presented in [20].

In addition to the cryostat described here for microwave measurements, we also constructed a modified instrument for studying the spectrum of the recombination radiation from Ge at temperatures of 0.5-1.5°K. The apparatus is shown in Fig. 5. It features certain characteristics dictated by the optical nature of the experiments. First, in order to increase the apertures, the He^3, He^4, and nitrogen Dewars have large openings, reaching 10 cm in the outer vessel. Second, there is a special aligning system for placing the sample exactly on the optical axis of the apparatus. Alignment is accomplished by moving the holder (2) to which the sample (1) is attached by means of the bellows (6) which is controlled by three thin-walled stainless steel tubes. The tubes are moved by screws with bellows sealed to the upper flange of the He^3 chamber (14).

The Ge crystal is greased (Apiezon) to the copper holder (2) which has a large surface area (about 25 cm^2). The surface area was obtained by soldering to a 2 × 2 cm copper plate about 200 fine copper wires 0.05 mm in diameter and about 100 mm long.

Experiments at temperatures below 1°K are difficult because of the poor thermal exchange between the liquid helium and the solid body. The main obstacle to reaching the bath temperature is the thermal resistance of Kapitsa [21] which appears at the boundary between two media because of a difference in acoustic impedance. Kapitsa showed experimentally that the boundary layer resistance R = $\Delta T/Q$, which is proportional to T^{-3}, causes a temperature discontinuity ΔT at the solid—liquid interface (Q is the heat flux across the boundary). This effect was first observed in the superfluidity of He^4 and later in He^3. The size of the Kapitsa discontinuity depends on the material and the surface condition, its purity, internal mechanical stresses, type of treatment, and so on. Detailed studies of the Kapitsa discontinuity are found elsewhere [22-25].

It is clear that the initial sample temperature is very important when a laser is used to excite Ge at temperatures below 1°K because even a small amount of incident power may cause warming. For example, let us calculate ΔT for a 10 × 5 × 0.7 mm Ge sample at 0.5°K. The sample is assumed to be completely immersed in He^3 and the laser delivers 5 mW continuous power. We shall use the following expression for the Kapitsa resistance [24]:

$$R = \sigma \Delta T/W = 50/T^3 \, \text{cm}^2\text{-deg/W}.$$

Here σ is the surface in contact with the liquid bath, T is the He^3 temperature, and W is the input power. We find that

$$\Delta T = \frac{W}{\sigma} \frac{50}{T^3} = \frac{5 \cdot 10^{-3} \cdot 50}{1 \cdot 0.125} = 2°.$$

This calculation indicates that small samples will be significantly warmed under our experimental conditions.

However, the heat transfer between solids is quite efficient at low temperatures. For example [25], the limiting resistance between two dissimilar metal surfaces is

$$R = 0.8/T^3 \, \text{cm}^2\text{-deg/W}.$$

Let us use this equation to calculate the temperature change at the Ge—copper boundary, assuming good mechanical contact between the materials. For T = 0.5°K, σ = 0.5 cm², and W = 5 mW we find

$$\Delta T = \frac{5 \cdot 10^{-3} \cdot 0.8}{0.5 \cdot 0.125} = 0.06°.$$

Thus, if the sample is well attached to the copper substrate, the temperature difference between the two pieces will be very small. Good thermal contact between the sample and substrate is ensured by a thin layer of silicone grease or Apiezon. It is even better to solder the sample to the holder, if possible. Now the Kapitsa discontinuity at the liquid interface is determined not by the sample surface but by the substrate surface which can be made as large as necessary.

The above considerations explain the various devices which we used in these systems to avoid heating the sample, including the radiator on the lower section of the copper waveguide in the microwave version of the cryostat and the bundle of fine wires in the optical version.

A number of special experiments were run to study the effects of heating. The sample temperature was controlled by either a carbon thermometer or a superconducting film; the results of these experiments are described in detail in [26]. We simply point out here that when the Ge is continuously excited the experimental values of ΔT agree well with calculations of the Kapitsa discontinuity if the sample and substrate are treated as a single unit. The experiments showed that in He³ at 0.5°K the warming was no greater than 0.1°K if the laser power is 5 mW.

We conclude by noting that the optical and microwave studies performed with the apparatus described here indicate that the instruments are very reliable. The He³ refrigerator allows us to hold the sample temperature at 0.5°K for a long time while the exciting laser is on, providing great possibilities for experimentation.

We wish to express our warmest thanks to K. N. Zinov'eva for useful advice and to V. M. Morkovin for building the devices.

LITERATURE CITED

1. N. Neganov, N. Borisov, and M. Liburg, Zh. Éskp. Teor. Fiz., 50:1445 (1966).
2. J. G. Daunt, Adv. Phys., 1:209 (1952).
3. K. N. Zinov'eva and V. P. Peshkov, Zh. Éksp. Teor. Fiz., 5:1025 (1957).
4. G. N. Mikhailova and V. B. Ginodman, Report of the Cryogenics Division of the Physics Institute, Academy of Sciences (1965).
5. V. P. Peshkov and N. I. Kondrat'ev, Prib. Tekh. Éksp., No. 4, p. 105 (1957).
6. L. V. Keldysh, in: Excitons in Semiconductors [in Russian], Nauka, Moscow (1971), p. 5.
7. M. Abkowitz and A. Honig, Rev. Sci. Instrum., 33:568 (1962).
8. I. Svare and G. Seidel, Phys. Rev., 134:A172 (1964).
9. I. Cowen, R. Spence, H. Van Till, and H. Weinstock, Rev. Sci. Instrum., 35:914 (1964).
10. T. L. Bohan and H. I. Stapleton, Rev. Sci. Instrum., 39:1707 (1968).
11. T. Ohyama, K. Murase, and E. Otsuka, J. Phys. Soc. Jpn., 29:912 (1970).
12. V. B. Ginodman, B. G. Zhurkin, V. F. Troitskii, and A. B. Fradkov, Prib. Tekh. Éksp., No. 2, p. 268 (1973).
13. A. B. Fradkov, G. N. Mikhailova, V. A. Milyaev, and S. I. Valyanskii, Kratk. Soobshch. Fiz., No. 7, p. 16 (1973).
14. Yu. M. Kholinov, Report of the Cryogenics Division, Physics Institute, Academy of Sciences (1969).
15. T. Halpern and H. Fritzsche, Rev. Sci. Instrum., 39:1336 (1968).
16. N. N. Mikhailov and A. Ya. Kaganovskii, Prib. Tekh. Éksp., No. 3, p. 194 (1961).

17.   G. K. White, Experimental Techniques in Low Temperature Physics, Oxford University Press (1958).
18.   G. N. Mikhailova, Zh. Tekh. Fiz., 41:800 (1971).
19.   T. R. Roberts and S. G. Sydoriak, Phys. Rev., 113:417 (1959).
20.   D. Waltom, Rev. Sci. Instrum., 37:6 (1966).
21.   P. L. Kapitsa, Zh. Éksp. Teor. Fiz., 11:1 (1941).
22.   G. L. Pollak, Rev. Mod. Phys., 41:48 (1969).
23.   N. S. Snyder, Cryogenics, 10:89 (1970).
24.   K. N. Zinov'eva, Zh. Éksp. Teor. Fiz., 60:2243 (1971).
25.   M. Suomi, A. Anderson, and B. Holmström, Physica, 38:67 (1968).
26.   A. A. Manenkov, G. N. Mikhailova, A. S. Seferov, and V. D. Chernetskii, Fiz. Tverd. Tela, 16:2719 (1974).

EFFECTS OF LASER HEATING ON GERMANIUM SAMPLES AT LIQUID HELIUM TEMPERATURE

A.A. Manenkov, G.N. Mikhailova, A.S. Seferov, and V.D. Chernetskii

The effect of warming Ge samples by pulsed and cw optical excitation with a YAG laser (1.06 μm) are studied at 1.3°K in He^3 and He^4. The thermal behavior of the samples is analyzed. A model is presented for calculating the thermal relaxation time. Experimental studies were performed using a superconducting Al bolometer evaporated onto the surface of the sample which, in turn, was attached to a massive heat sink. Under these conditions warming is observed when the energy of the exciting pulse E $\geq$ 5 · 10^{-6} J. For E = 10^{-4} J the warming did not exceed 0.13°K and lasted 10 μsec in a He^4 bath. Both the amount of warming and its duration (up to 25 μsec) were greater in He^3. Results obtained with continuous excitation of the Ge are in good agreement with calculations based on the Kapitsa discontinuity equation.

Laser radiation of various wavelengths and intensities is used to study the properties of semiconductors at low temperatures. Optical excitation produces nonequilibrium carriers and phonons in a crystal, their number and energy being dependent on the ratio of the semiconductor band gap to the photon energy hν and on the number of photons, i.e., the light intensity. When h$\nu \geq \Delta E_0$ electrons are promoted from the valence band to the conduction band, and in a time period much shorter than the photoelectron's lifetime the electron relaxes to the bottom of the conduction band, giving up its excess energy h$\nu - \Delta E_0$ to the lattice. As a result, the sample heats up and many of the electrical and thermal properties, which are temperature dependent, change in response. This effect can be very important at helium temperatures when a small amount of heating, of the order of tenths of a degree, can cause substantial errors in certain measured parameters, especially when they have a strong temperature dependence in the region of 1-4°K. A case in point is the intensity of recombination radiation from excitons and electron-hole droplets in Ge [1]. Thus, we found it necessary to perform a detailed study of the temperature behavior of samples placed in liquid helium and irradiated with optical radiation, both pulsed and cw.

In general the crystal temperature depends on the exciting power, the sample dimensions, the means of cooling, and the temperature of the medium in which it is found. An important and experimentally well-studied case is that of the optical excitation of a sample freely suspended in superfluid helium. Here, because of the difference in phonon velocities in the solid and liquid, there is a temperature discontinuity ΔT at the sample−liquid interface; this is called the Kapitsa discontinuity. The discontinuity is proportional to the heat current and inversely proportional to the sample area:

$$\Delta T = RW/\sigma. \qquad (1)$$

The proportionality coefficient R for He^4 has been measured in the range 0.5-1.5°K for a number of different materials and is equal to

$$R \simeq (40 \text{ to } 50) \; T^{-3} \; \text{cm}^2\text{-deg/W}. \qquad (2)$$

It is called the Kapitsa boundary resistance [2].

Let us estimate the heating of a Ge sample 10 × 5 × 0.7 mm in He^4 at 1.5°K when it is irradiated by a cw laser with 50 mW power:

$$\Delta T = \frac{50}{T^3} \frac{W}{\sigma} \simeq 0.6°,$$

Thus, under these conditions the heating can be a very large fraction of the sample temperature especially when the temperature is lowered and the optical power increased. To avoid such great heating it is essential to replace the solid−liquid contact with a solid−solid junction; i.e., the sample must be attached to a massive substrate with good thermal conductivity (the best choice is copper). The nonequilibrium phonons which heat the sample will then leave it much faster because the sound velocities in the two contacting pieces are more nearly equal.

No less important, but much more difficult to analyze, is the case of pulsed optical excitation. When the sample is irradiated with sufficiently short pulses one can estimate the heating at the time of the pulse by using an adiabatic approximation which ignores the heat flow to the helium bath:

$$Q = \int_{T_0}^{T_1} C(T) \, m \, dT. \qquad (3)$$

Here m is the sample mass,* C(T) is the specific heat, and Q is the absorbed energy. We will show below that "sufficiently short" pulses are those whose durations $\tau \ll 10^{-6}$ sec.

At low temperatures (T $\ll$ Θ_D) the specific heat of Ge follows the law [5]

$$C = [27 \, (T/\Theta_D)^3 + 3.6 \cdot 10^{-7} \, T] \; \text{J/(g-deg)}, \qquad (4)$$

where Θ_D = 400°K is the Debye temperature.

Table 1 presents results for the calculated heating of a 0.15-g-mass Ge sample based on Eqs. (3) and (4) for certain experimental conditions.

We see that relatively small amounts of laser energy will cause noticeable heating, expecially at low temperatures.

Another important factor which must be included in considerations of the thermal behavior of irradiated samples is the time needed to establish thermal equilibrium with the surrounding medium following the pulsed excitation. It is evident that valid measurements of many temperature-dependent quantities will be very difficult during the equilibrating period. In order to deal with the cooling dynamics one must use the nonstationary thermal conductivity equation

$$dT/dt = a\nabla^2 T, \qquad (5)$$

---

*Although carriers are produced only in a thin surface layer (the optical absorption coefficient of Ge is about 10^4 cm^{-1} at 1.06 μm), they rapidly diffuse [4] uniformly throughout a sample 1 mm thick in 1 μsec.

TABLE 1

Bath temperature, °K	Laser pulse, J	Heating, °K
4.2	10^{-4}	4.8
	10^{-5}	1.3
	10^{-6}	0.2
2.5	10^{-4}	6.4
	10^{-5}	2.6
	10^{-6}	0.7
1.3	10^{-4}	7.5
	10^{-5}	3.7
	10^{-6}	1.5

where T is the temperature, t is the time, and $a = \lambda/c\rho$ is the thermal diffusivity (λ is the thermal conductivity, c is the specific heat, and ρ is the density).

To simplify the analysis we will calculate the time to thermal equilibrium of a Ge crystal with a helium bath using a model in which a body heated to an initial temperature T_i is immersed in a medium with temperature T_0 (boundary conditions of type III). The solution to this problem takes the form [6]

$$\frac{T - T_0}{T_i - T_0} = \sum_{n=1}^{\infty} A_n \cos\left[\mu_n(1 - \eta)\right] e^{-\mu_n^2 F_0}, \tag{6}$$

where

$$\cot \mu_n = \frac{1}{Bi}\mu_n, \qquad A_n = (-1)^{n+1}\frac{2\,Bi\sqrt{\mu_n^2 + Bi^2}}{\mu_n(\mu_n^2 + Bi^2 + Bi)},$$

Bi = $h\alpha/\lambda$ is the Biot number, $F_0 = at/h^2$ is the Fourier number, η is the dimensionless coordinate or thickness parameter, μ_n is a dimensionless coefficient, h is one-half the plate thickness, and α is the heat-transfer coefficient.

As an example we have calculated the cooling time of a Ge sample 2h = 1 mm thick from $T_i = 1.8°K$ to T = 1.4°K when it is immersed in a He^4 bath at $T_0 = 1.3°K$. The thermal conductivity of Ge at 2°K is $\lambda = 1$ W/(cm-deg) [7], and a is taken at an intermediate temperature of 1.5°K: $a = 1.2 \cdot 10^5$ cm^2/sec. The heat-transfer coefficient in He^4 at 1.3°K is $\alpha = 0.2$ W/(cm^2-deg) [8]. Thus we have Bi = 10^{-2} and η = x/h = 1 (the heating process is computed at the sample center). The temperature parameter $(T - T_0)/(T_i - T_0) = 0.2$, and $\mu_1 = 10^{-1}$. Keeping just the first term (n = 1) of the series in Eq. (6) and substituting the values of μ_1 and Bi into A_1 we find (assuming that $A_1 \approx 1$)

$$\frac{T - T_0}{T_i - T_0} = e^{-\mu_1^2 F_0}. \tag{7}$$

Using the expression for F_0 we can obtain an equation for the time needed to cool the sample to temperature T, within the scope of our model:

$$t = \frac{h^2}{\mu_1^2 a} \ln\left(A_1 \frac{T_i - T_0}{T - T_0}\right). \tag{8}$$

By substituting the parameters selected for our example into Eq. (8) we obtain

$$t \simeq 3.3\ \mu\text{sec.}$$

It is of interest to inquire into how much the sample cools in a period $\tau = 10^{-7}$ sec, which is the duration of the pulse used by our laser. Solving Eq. (8) now for T instead of time using the same values of T_0 and T_i we find T = 1.78°K. This means that in time τ the sample temperature drops only 0.02°K (T_i = 1.8°K) due to heat transfer to the helium. It is therefore clear that, in order to use the adiabatic approximation to estimate the amount of sample warming, one requires laser pulses of this length so that, while the pulse is on, the temperature drop of the sample will be much smaller than the temperature difference due to the instantaneous transfer of heat between the helium bath and sample. In our case this latter difference is $T_i - T_0$ = 0.5°K $\gg$ 0.02°K, a result which validates our use of the adiabatic approximation above.

The Kapitsa thermal boundary resistance R is found from Eq. (1), just as for the case of cw excitation, where, however, ΔT and W are taken to be time dependent; R determines the value of the heat-transfer coefficient α. Note that Eq. (2) is satisfied in He^3 for temperatures lower than 0.5°K, but when T is greater than 1°K the heat-transfer coefficient of He^3 is about 100 times smaller than in He^4 [3].

In order to directly measure the amount and duration of pulsed warming of a Ge sample due to optical excitation we require a thermometer which has a rapid response time, high sensitivity, and good thermal contact with the sample. With these requirements on the thermometer, the best approach to the problem is a thin, superconducting film on the sample surface but electrically isolated from it. Near the superconducting transition point the large value of dR/dT and small specific heat of the film makes it possible to detect rapid temperature changes due to a laser pulse. The choice of superconductor used is mainly determined by the operating temperature. A lead superconducting bolometer (T_c = 2.19°K) was used for similar purposes in [9, 10].

At temperatures in the range of 1.3-4.2°K, which are reached by pumping on He^4, one can use such superconductors as Al (T_c = 1.196°K), In (T_c = 3.403°K), Ta (T_c = 4.483°K), Sn (T_c = 3.722°K), and other metals and alloys. In our experiments we used pure aluminum because its critical temperature lies in the temperature range where one expects the greatest relative warming.

A bolometer, which is a heat-sensing device, was fabricated as follows. On one face of the $10 \times 5 \times 0.7$ mm Ge sample we evaporated a 1000 Å thick film of SiO. This film was of such thickness as not to materially affect the thermal contact of the bolometer with the crystal but was still able to electrically isolate the bolometer from the sample. Then, special masks and vacuum evaporation were used to lay down four parallel gold pads for current and voltage contacts; these pads were about 1 μm thick. Finally, perpendicular to these pads was evaporated a film 300 Å thick and 0.5 mm wide. The dimensions of the bolometer film were selected based on contradictory requirements, that of increasing sensitivity by decreasing cross section (which increases resistance) and decreasing mechanical strength in the process. Electrical leads were attached to the gold pads with conductive silver glue.

Figure 1 shows the layout of the sample with the bolometer and cryostat. To improve thermal contact between the germanium sample and the helium bath the sample was attached with Apiezon grease to the copper heat sink which had an expanded area (about 20 cm^2). The laser beam was focused onto the opposite side of the sample through an aperture in the heat sink. We used a Nd:YAG laser (1.06 μm) which operates in both pulsed and cw modes. In the pulsed mode the laser pulses were τ_i = 0.1 μsec long with a repetition rate of 100 Hz and a maximum of 10^{-4} J per pulse.

The bolometer resistance was measured potentiometrically with a current of 1 mA. At room temperature its resistance was about 35 Ω. Since the method of recording pulsed warming is based on the transition of the film from the superconducting to the normal state, which is evidenced by a voltage pulse, it is necessary to calibrate the temperature dependence of the

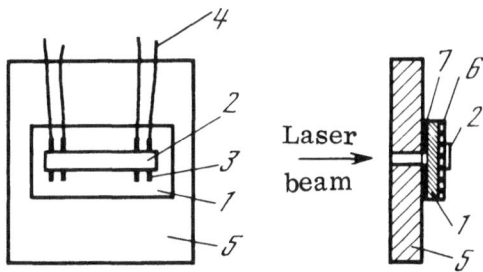

Fig. 1. Layout of sample and bolometer. 1, Ge sample; 2, superconducting Al bolometer; 3, Au electrical pads; 4, current and voltage leads; 5, copper heat sink (1 mm thick, 20 cm² area); 6, 1000 Å thick SiO film; 7, Apiezon grease.

bolometer resistance in the neighborhood of the superconducting transition in order to uniquely identify the voltage pulse amplitude with the bolometer temperature. The calibration is based on the pressure of the saturated He^3 vapor. The calibration curves, shown in Fig. 2, have good reproducibility and show the width and hysteresis and increase in T_c which are all characteristic of thin films.

These measurements require a current which does not cause a substantial thermal effect on the bolometer when the resistive state is reached, but which still provides enough sensitivity for the measurement. Short current pulses are frequently used to eliminate Joule heating of films when one is obtaining the transition curves due to magnetic effects of current. These are the so-called isothermal transition curves in which the parasitic thermal effect is reduced to a minimum [11]. Similar effects can occur when dealing with the thermal transition if the current through the film, which is to serve only as an indicator of the resistive state, causes additional heating of the sample. We shall treat this effect in more detail.

As the temperature is raised in a superconducting film, there is a moment when resistance appears in some small region. The density of the measuring current flowing around this nonsuperconducting center increases because the effective cross section is thereby reduced. This causes new portions of the film to go into the normal state, but the normal region does

Fig. 2. Superconducting transition curves for Al bolometer. 1) $T_{\downarrow c} = 1.397°K$; 2) $T_{\uparrow c} = 1.470°K$.

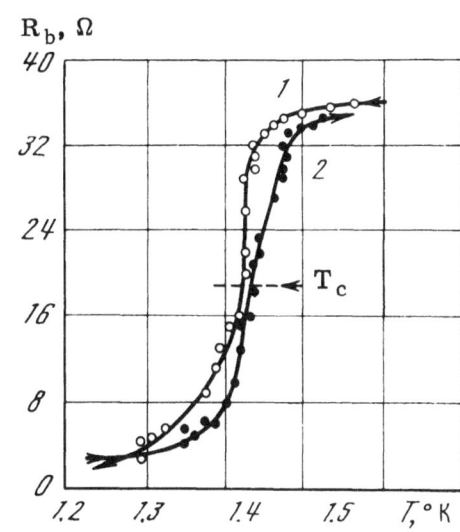

not cover the film's cross section by a thin, channel-like jumper. This moment marks the origin of the superconducting transition curve because resistance begins to appear in the film and Joule heat is released. The heat causes the normal region to expand, and the resistance increases. An avalanche process arises which is called thermal propagation, the spontaneous increase in the nonsuperconducting region even at constant current and temperature, right up to the total, discontinuous appearance of resistance. As the film is cooled this discontinuity appears at a lower temperature because the large initial size of the normal region ensures the release of a considerable amount of Joule heat. Therefore, if thermal effects exert a large influence on the transition, the transition will exhibit considerable hysteresis.

Thermal propagation appears when the thermal energy released because of an increase in the size of the normal region is greater than the heat transferred from the film to the surroundings. Thus, the thermal conductivity of the substrate and the cryogen have a strong effect on the acuity of the effect.

Thermal propagation was apparently not a factor in our constant-current measurements because, as the experimental transition curves show (Fig. 2), the region of discontinuous resistance change is relatively small and the hysteresis in this neighborhood is rather slight (about 10% over the entire transition region). The thin-film Al bolometer sensitivity was $dR/dT = 420\ \Omega/\deg$ in the region of greatest slope (near $T = T_c$), which enabled us to measure pulsed warming in the amount $\Delta T = 0.01°$. The width of the superconducting transition was $0.2°$, which is very large in comparison with similar transitions in bulk superconductors (for instance, in pure tin the transition is about $0.001°$ wide). Therefore, we could detect warming in the region $0.01 \le \Delta T \le 0.2°$. If the heating exceeded $0.2°$ the amplitude of the signal was limited by saturation of the transition curve to the normal state, but the amount of warming could still be estimated nevertheless from the duration of the plateau.

We performed studies of the heating and cooling kinetics of the sample using different laser powers. In the pulsed mode the signal from the potential leads of the bolometer was applied directly to the input of an oscilloscope and the amplitude and duration of the heating pulses could be taken directly from the scope screen. Figure 3 shows the shape of the bolometer voltage pulses for various excitation levels. Figure 4 shows the time dependence of the sample temperature for an excitation level $E = 5 \cdot 10^{-5}$ J per pulse and $T = 1.3°$K, obtained using the calibration curve. We see that the trailing edge of the pulse, which characterizes

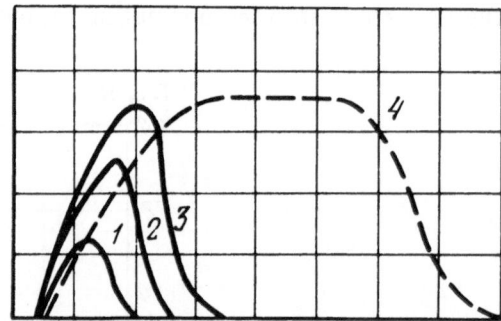

Fig. 3. Shape of the bolometer voltage pulse for different laser energy pulses. E, joules: 1) $2 \cdot 10^{-5}$; 2) $3 \cdot 10^{-5}$; 3) $5 \cdot 10^{-5}$ in a He⁴ bath; 4) $4 \cdot 10^{-5}$ in a He³ bath. The limit to the signal amplitude is due to nonlinearities in the function $R(T)$. Abscissa is $2\ \mu\text{sec/div.}$, ordinate is 5 mV/div. Current through bolometer is 1 mA.

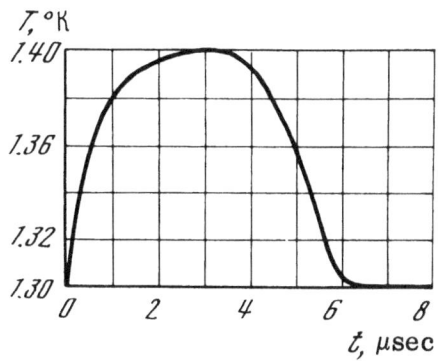

Fig. 4. Time dependence of the sample temperature in He⁴ for a laser pulse. $E = 5 \cdot 10^{-5}$ J, $T = 1.3°K$.

the sample cooling process, is a few microseconds long, in good agreement with the sample cooling time (about 3 μsec) calculated from Eq. (7) using conditions similar to those of the experiment.

Based on these results for pulsed heating in He³ and He⁴ at 1.3°K, we have constructed graphs of the dependence of the warming amplitude and duration on the excitation level; these are shown in Figs. 5 and 6. We see that with our experimental conditions (T = 1.3°K, 0.15-g sample fastened to a heat sink) warming begins when the laser pulse energy exceeds $5 \cdot 10^{-6}$ J. The maximum detected heating in He⁴ is 0.13° for $E = 10^{-4}$ J per pulse. Because of its smaller

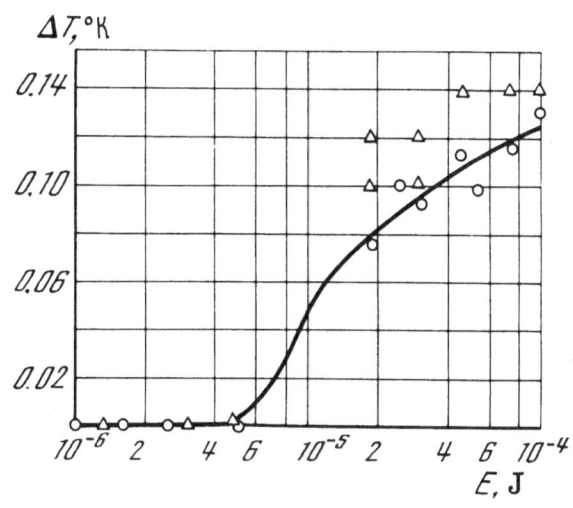

Fig. 5. Dependence of the sample warming pulse on the laser energy pulse at 1.3°K in He⁴ (triangles) and in He³ (circles). At high excitation levels in He³ we could not measure the warming which took the bolometer beyond the limits of the transition curve into the normal state.

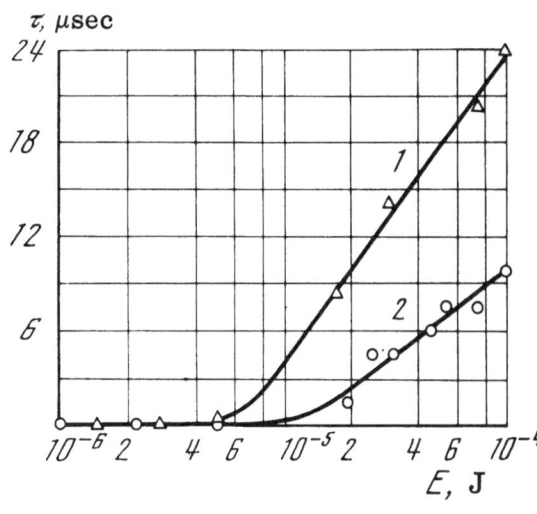

Fig. 6. Dependence of the warming pulse duration (at 0.1 times the amplitude) on the laser pulse energy. 1) in He³; 2) in He⁴.

thermal transfer coefficient, it was not possible to determine the amount of heating in He^3 when E > 10^{-5} J per pulse because the data points fell in the saturation region of the super-conducting transition curve, and the bolometer sensitivity goes to zero. Figure 6 shows the dependence of the warming pulse durations on the energy of the exciting pulse. The maximum duration in He^4 is 10 μsec for E = 10^{-4} J per pulse, and it is much larger in He^3 (about 25 μsec).

Let us now turn to the fact that the calculated results in the table for the warming of a sample freely suspended in helium are as much as two orders of magnitude greater than the experimental values of ΔT. Because the leading edge of the bolometer signal lasts much longer than one would expect by calculating the thermal propagation time in the sample, the discrepancy can be explained by the thermal inertia of the bolometer [12]. If the bolometer were triggered a few microseconds following the laser pulse, the sample would already be cooled and we would record some intermediate amount of warming as the maximum warming. Thus, because of the inertia in the bolometer, we cannot in principle observe the heating predicted by calculations with the adiabatic approximation (Table 1). Nevertheless, we can say with assurance that after 2-3 μsec following the laser pulse of 10^{-4} J the sample temperature differs from the He^4 bath temperature by no more than 0.2°K.

We conducted experiments with this same bolometer using continuous excitation. In He^4 at 1.3°K the heating did not exceed 0.1°K when the laser power was 50 mW. The temperatures were measured in the steady-state mode using the calibration curve. As for measurements in He^3, they were performed at 0.5°K using a different method* because with this excitation power the steady-state temperature was outside the range of the superconducting transition. A resistance thermometer attached to the copper heat sink showed a warming of 0.7°K. Both experimental quantities are in good agreement with Eq. (1) if the sample and heat sink are treated as a single unit and one treats the heating of the entire system relative to the helium bath. As mentioned above, the Kapitsa discontinuity between solids at low temperatures is much smaller than that for a solid−liquid interface. Data indicate [14] that the boundary resistance for a metal−metal contact is

$$R = \sigma \Delta T/W = 0.8/T^3 \quad \text{cm}^2 \cdot \text{deg/W}.$$

Calculation indicates that the temperature difference between sample and heat sink is insignificant when their thermal contact is good, while the warming of a sample with a heat sink relative to the helium bath is determined mainly by the surface of the heat sink, which can be made quite large.

In conclusion, we list the basic results of our study.

1. The model presented here for calculating the thermal relaxation time after the sample is illuminated by a laser pulse gives good agreement with experiment.
2. The warming of a Ge crystal attached to a large-area heat sink submerged in He^4 at 1.3°K is 0.13°K with a warming time τ = 10 μsec when the laser energy is E = 10^{-4} J per pulse.
3. No warming is observed for either He^4 or He^3 at 1.3°K when the excitation energy is E < 5 · 10^{-6} J per pulse.
4. Experiments conducted with He^3 indicate warming and thermal relaxation times in the sample which are much larger than those in He^4.
5. When Ge is excited with continuous laser radiation the experimental values of ΔT agree well with the values calculated with the Kapitsa discontinuity equation.

---

* The device in which these measurements were performed is described in [13].

We wish to thank V. A. Milyaev, É. A. Tishchenko, V. F. Troitskii, and A. B. Fradkov for useful discussions, Ya. G. Ponomarev for assistance in fabricating the bolometers, and V. A. Sanina for help in conducting the experiments.

LITERATURE CITED

1. L. V. Keldysh, in: Excitons in Semiconductors [in Russian], Nauka, Moscow (1971), p. 5.
2. P. L. Kapitsa, Zh. Éksp. Teor. Fiz., 11:1 (1941).
3. K. N. Zinov'eva, Zh. Éksp. Teor. Fiz., 60:2243 (1971).
4. B. N. Novikov, E. F. Gross, and M. A. Drygin, Pis'ma Zh. Éksp. Teor. Fiz., 8:15 (1963).
5. H. M. Rosenberg,, Proc. Phys. Soc., A67:837 (1967).
6. A. I. Pekhovich and V. M. Zhidkikh, Calculations of Thermal Properties of Solids [in Russian], Énergiya, Leningrad (1968), pp. 157, 269.
7. Handbook of Physical and Technical Foundations of Cryogenics [in Russian], M. P. Malkova, ed., Énergiya, Moscow (1973), p. 133.
8. N. S. Snyder, Cryogenics, 10:89 (1970).
9. J. C. Hensel and T. G. Phillips, Proceedings of the Eleventh International Conference on the Physics of Semiconductors, Warsaw, 1972, p. 671.
10. M. Gurny and M. Gliksman, Solid State Commun., 7:11 (1972).
11. D. Bremer, Superconducting Devices [Russian translation], Mir, Moscow (1964), p. 34
12. R. J. von Gutfeld, in: Physical Acoustics, W. P. Mason, ed., Vol. 5, Academic Press, New York (1968), p. 233.
13. T. I. Galkina, V. A. Milyaev, G. N. Mikhailova, and N. A. Penin, Pis'ma Zh. Éksp. Teor. Fiz., 18:99 (1973).
14. M. Suomi, A. Anderson, and B. Holmström, Physica, 38:67 (1968).

CYCLOTRON RESONANCE AND RADIATIVE RECOMBINATION IN PURE Ge WITH LASER EXCITATION

V.P. Aksenov, N.B. Volkov, B.G. Zhurkin, and I.G. Maksimchuk

We have measured the kinetics of cyclotron resonance (CR) and radiative recombination (RR) in pure germanium at 1.8°K using high levels of optical excitation. It is found that the free carrier lifetime as determined from the decay of the CR line intensity can be described as the sum of two exponentials: $\tau_{CR_1} = 72 \pm 5$ μsec and $\tau_{CR_2} = 36 \pm 2$ μsec. The free carrier lifetime obtained from the decay of the LA(709) line is $\tau_{RR} = 36 \pm 2$ μsec. Using the rate equation for the system of free carriers, excitons, and electron-hole droplets, and noting that the mean free path of Auger electrons leaving the EHD is longer than the EHD radius, the ratio of τ_{CR} to τ_{RR} is obtained. When the free carrier density $n_c > 10^{11}$ cm^{-3}, $\tau_{CR}/\tau_{RR} \simeq 2$ while when $n_c < 10^{11}$ cm^{-3}, $\tau_{CR}/\tau_{RR} \simeq 1$, in agreement with the experimental data.

Gladkov et al. [1] performed the first simultaneous studies of cyclotron resonance (CR) and radiative recombination (RR) in pure Ge at temperatures of 1.8-4.2°K with high levels of optical excitation. It was determined that the ratio of the free carrier lifetime, determined from the decay time τ_{CR} of the cyclotron resonance line, to the radiative recombination time τ_{RR} of carriers in electron-hole droplets, as determined from the decay of the LA(709) line, is $\tau_{CR}/\tau_{RR} \simeq 2$. On the other hand, Hensel et al. [2] found that $\tau_{CR}/\tau_{RR} \simeq 1$. This discrepancy in the experimental results has motivated us to once again study the kinetics of both CR and RR in pure germanium.

The work was performed using pure germanium samples containing 10^{11} cm^{-3} residual impurities. The sample dimensions were $4 \times 4 \times 0.3$ mm. The kinetics of the CR line and the recombination radiation were detected simultaneously by using pulsed optical excitation and stroboscopic integration (Fig. 1). The excitation pulse lasted 3 μsec and the strobe pulse was 0.5 μsec. All measurements were performed at 1.8°K where the recombination radiation spectrum contains only the LA(709) line from recombination in the electron-hole droplets.

EXPERIMENTAL RESULTS

Figure 2 shows the dependence of the CR and RR intensities on the time delay of the strobe pulse. The relaxation of the CR intensity is the same for the two types of carriers. The slow part of the relaxation is well described by an exponential with a time constant of $\tau_{CR_1} = 72 \pm 5$ μsec and the fast portion corresponds to an exponential with time constant $\tau_{CR_2} = 36 \pm 2$ μsec. The location of the inflection point A' depends on the level of excitation. As the excitation level is reduced the inflection point shifts toward shorter delays and finally disappears. Only the exponential with $\tau_{CR_2} = 36$ μsec remains. The free carrier density at the time $t_{A'}$ is $n = 10^{10}$-10^{11} cm^{-3}. The lifetime for radiative recombination from the electron-hole droplets is 36 ± 2 μsec.

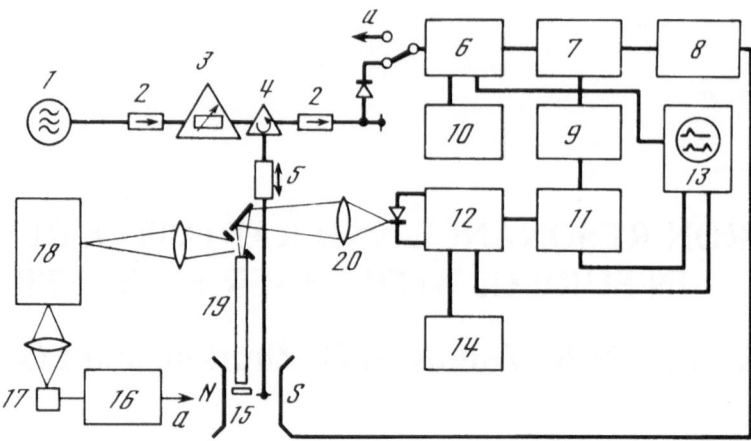

Fig. 1. Block diagram of apparatus used to study CR and RR. 1, microwave generator; 2, ferrite rectifier; 3, attenuator; 4, ferrite circulator; 5, impedance transformer; 6, wide-band amplifier; 7, strobe integrator; 8, XY recorder (PDS-021); 9, strobe pulse generator; 10, power supply; 11, GaAs laser trigger; 12, current pulse generator for GaAs laser; 13, S1-17 oscilloscope; 14, power supply; 15, Ge sample; 16, wide-band preamplifier; 17, FD-111 photodiode; 18, MDR-3 monochromator; 19, light pipe; 20, GaAs laser.

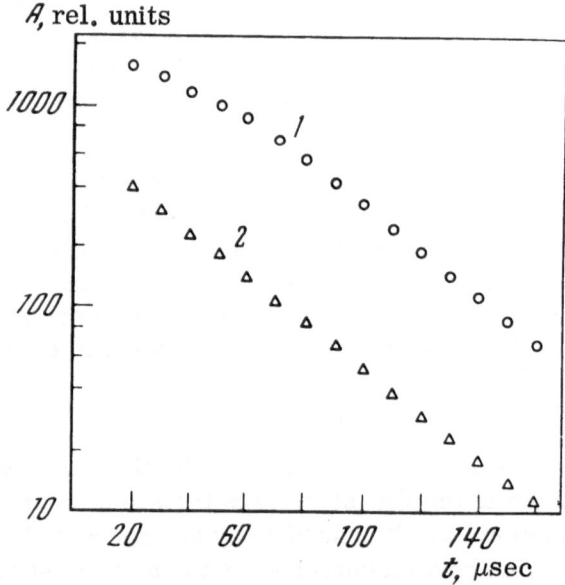

Fig. 2. CR and RR amplitudes (1 and 2 respectively) as functions of the strobe pulse delay for an excitation power corresponding to an electron-hole pair density of 10^{16} cm^{-3}; temperature is 1.8°K.

DISCUSSION

Betzler et al. [3] have calculated the fraction of Auger processes in carrier recombination in droplets; it amounts to about 75%. These data can be used to interpret our results in terms of the "drop model."

Consider the equation governing the time dependence of the free carrier density in a system of excitons, droplets, and free carriers following the light pulse:

$$dn_c/dt = -\sigma_{eh}v_T n_c^2 - n_c/\tau_c + A + B, \qquad (1)$$

where n_c is the free carrier density σ_{eh} is the cross section for binding an electron and hole into an exciton, τ_c is the free carrier lifetime for recombination, and v_T is the carrier thermal velocity.

The quantity A, which describes exciton dissociation, will be neglected; B describes the rate at which Auger electrons are produced from droplets:

$$B = \begin{cases} \dfrac{N}{V\tau_A} \propto N, & \Delta \gg R, \\[2ex] \dfrac{3\Delta}{R}\dfrac{N}{V\tau_A} \propto N^{2/3}, & \Delta \ll R, \end{cases} \qquad (2)$$

where N is the total number of particles in the droplet; $N \propto \exp(-t/\tau_{RR})$, where $\tau_{RR} = 36 \pm 2$ μsec from measurements of radiative recombination, τ_A is the time describing the rate of Auger recombination, R is the droplet radius, V is the sample volume, and Δ is the effective mean free path length of an Auger electron.

Equation (1) is quasistationary with respect to n_c and can be rewritten as

$$\sigma_{eh}v_T n_c^2 + n_c/\tau_c \simeq B. \qquad (3)$$

Let us consider a number of limiting cases:

1. When $n_c > 10^{11}$ cm^{-3} and $\sigma_{eh}v_T n_c^2 \gg n_c/\tau_c$, we find from Eq. (3) that $n_c \simeq (B/\sigma_{eh}v_T)^{1/2}$, so that

$$n_c \propto \begin{cases} N^{1/2} \propto \exp(-t/2\tau_{RR}), & \Delta \gg R, & \tau_{CR} = 2\tau_{RR} = 72 \ \mu\text{sec}, \\[1ex] N^{1/3} \propto \exp(-t/3\tau_{RR}), & \Delta \ll R, & \tau_{CR} \simeq 3\tau_{RR} = 108 \ \mu\text{sec}. \end{cases}$$

2. When $n_c < 10^{11}$ cm^{-3}, $n_c = B\tau_c$:

$$n_c \propto \begin{cases} N \propto \exp(-t/\tau_{RR}), & \Delta \gg R, & \tau_{CR} = \tau_{RR} = 36 \ \mu\text{sec}, \\[1ex] N^{2/3} \propto \exp(-2t/3\tau_{RR}), & \Delta \ll R, & \tau_{CR} \simeq 1.5\tau_{RR} = 54 \ \mu\text{sec}. \end{cases}$$

Microwave absorption studies [4] give $\tau \simeq 54$ μsec, which the authors explain in terms of generating free carriers from the droplet surface. Our results indicate that free carriers are generated throughout the entire volume of the drop, i.e., $R \leq \Delta$.

LITERATURE CITED

1. P. S. Gladkov, B. G. Zhurkin, and N. A. Kazakov, Tr. Mosk. Fiz. Tekh. Inst., Obshch. Mol. Fiz., pp. 1-9 (1973).
2. J. C. Hensel, T. G. Phillips, and T. M. Rice, Phys. Rev. Lett., 30:227 (1973).
3. K. Betzler, B. G. Zhurkin, and A. L. Karuzskii, Solid State Commun., 17:577 (1957).
4. A. A. Manenkov, V. A. Milyaev, G. N. Mikhailova, V. A. Sanina, and A. S. Seferov, Zh. Éksp. Teor. Fiz., 70:695 (1976).